SpringerBriefs in Electrical and Computer Engineering

For further volumes:
http://www.springer.com/series/10059

Carlo Dal Mutto • Pietro Zanuttigh
Guido M. Cortelazzo

Time-of-Flight Cameras and Microsoft Kinect™

 Springer

Carlo Dal Mutto
University of Padova
Padova, Italy

Pietro Zanuttigh
University of Padova
Padova, Italy

Guido M. Cortelazzo
University of Padova
Padova, Italy

ISSN 2191-8112　　　　ISSN 2191-8120 (electronic)
ISBN 978-1-4614-3806-9　　ISBN 978-1-4614-3807-6 (eBook)
DOI 10.1007/978-1-4614-3807-6
Springer New York Heidelberg Dordrecht London

Library of Congress Control Number: 2012935106

Springer is part of Springer Science+Business Media (www.springer.com)

To my father Umberto, who has continuously stimulated my interest for research

To Marco, who left us too early leaving many beautiful remembrances and above all the dawn of a new small life

To my father Gino, professional sculptor, to whom I owe all my work about 3D

Preface

This book originates from the activity on 3D data processing and compression of still and dynamic scenes of the Multimedia Technology and Telecommunications Laboratory (LTTM) of the Department of Information Engineering of the University of Padova, which brought us to use Time-of-Flight cameras since 2008 and lately the Microsoft Kinect^{TM} depth measurement instrument. 3D data acquisition and visualization are topics of computer vision and computer graphics interest with several aspects relevant also for the telecommunication community active on multimedia.

This book treats a number of methodological questions essentially concerning the best usage of the 3D data made available by ToF cameras and Kinect^{TM}, within a general approach valid independently of the specific depth measurement instrument employed.

The practical exemplification of the results with actual data given in this book does not only confirm the effectiveness of the presented methods, but it also clarifies them and gives the reader a sense for their concrete possibilities.

The reader will note that the ToF camera data presented in this book are obtained by the Mesa SR4000. This is because Mesa Imaging kindly made available their ToF camera and their guidance for its best usage in the experiments of this book. The products of any other ToF cameras manufacturers could be equivalently used.

This book has become what it is thanks to the contributions of a number of people which we would like to acknowledge. First of all, we are very grateful to all the students who lately worked in the LTTM laboratory and among them Enrico Cappelletto, Mauro Donadeo, Marco Fraccaro, Arrigo Guizzo, Giulio Marin, Alessio Marzo, Claudio Paolini, Mauro Tubiana, Lucio Bezze, Fabrizio Zanatta and in particular Fabio Dominio, who deserves our special acknowledgment for his great help in the experiments reported in this book. We would also like to thank Davide Cerato and Giuliano Pasqualotto (with 3Deverywhere) who worked on a number of data fusion topics with us, Simone Carmignato and Stefano Pozza (with DIMEG of the University of Padova) for their help on the metrological analysis of depth cameras, Gerard Dahlman and Tierry Oggier for the great collaboration we received from Mesa Imaging, Arrigo Benedetti (now with Microsoft, formerly with Canesta) and

Abbas Rafii (formerly with Canesta) for the organization of student exchanges between Canesta and our laboratory.

This book benefited from the discussions and the supportive attitude of a number of colleagues among which we would like to recall David Stoppa and Fabio Remondino (with FBK), Roberto Manduchi (with U.C.S.C.), Luciano Gamberini (with the Psychology Department of the University of Padova), Marco Andreetto (with Google), Tim Drotz (with SoftKinetic), Eran Ripple (with Primesense) and Radu B. Rusu (with Willow Garage). We must also thank Ruigang Yang, Lian Wang, Ryan Crabb, Jiejie Zhu, James E. Davis, Zhigeng Pan, Chenxi Zhang, Cha Zhang, Timo Kahlmann and Hilmer Ingesand for the figures showing the results of their research activity in the book.

Padova,   *Carlo Dal Mutto*
December 2011   *Pietro Zanuttigh*
                                                                                  Guido Maria Cortelazzo

Contents

Chapter 1
Introduction

The acquisition of the geometry description of a dynamic scene has always been a very challenging task which required state-of-the-art technology and instrumentation only affordable by research labs or major companies until not too long ago. The recent arrival on the market of matricial Time-of-Flight range cameras (or simply *ToF cameras* in this book) and the even more recent introduction of the Microsoft Kinect^{TM} range camera (or simply *Kinect* in the sequel), accompanied by massive sales of this product, has made widely available depth data streams at video rate of generic still or dynamic scenes. A truly amazing possibility considering that until now streaming at such rates was possible only with standard image sequences. In the computer graphics, computer vision and image processing communities a large interest arose for these devices and questions of the following kind have become common: "What is a ToF camera?", "How does the Kinect^{TM} work?", "Are there ways to improve the low resolution and high noise characteristics of ToF cameras data?", "How far can I go with the depth data provided by a 150 Euros Kinect^{TM} with respect to those provided by a few thousand Euros ToF camera?". The motivation of this book can be summarized by saying that it tries to address this kind of questions from a data user (and not from a device technology developer) point of view.

This book firstly describes the technology behind ToF cameras and the Kinect^{TM} and then focuses on how to make the best from the data produced by ToF cameras and Kinect^{TM}, i.e., on the data processing methods best suited to depth data. The depth nature of the data is used as a leverage to present approaches as much device-independent as possible. In this sense the book refers as often as possible to *depth cameras* and it makes the distinction between ToF cameras and Kinect^{TM} only when necessary.

The book perspective centered on the depth nature of the data, rather than on the devices, gives a common framework not only suited to both ToF and Kinect^{TM} current data, but also ready to be applied to the new devices of these families that will reach the market in the next years.

Although ToF cameras and Kinect^{TM} as depth cameras, i.e, as providers of depth data, are functionally equivalent, it is also important to remember that there exist fundamental technological differences between them which cannot be ignored.

The synopsis of distance measurement methods of Figure 1.1, derived from [1], offers a good framework to introduce such differences. For the purposes of this book it suffices to note that the reflective optical methods of Figure 1.1 are typically classified into *passive* and *active*. Passive range sensing refers to 3D distance measurement by way of radiation (typically, but not necessarily, in the visible spectrum) already present in the scene, and stereo-vision systems are a classical example of this family of methods; active sensing refers, instead, to 3D distance measurement obtained by projecting in the scene some form of radiation as made, for instance, by ToF cameras and by light coding systems of which the Kinect^{TM} is a special case.

The operation of ToF cameras and Kinect^{TM} involves a number of different concepts about ToF sensors, imaging systems and computer vision recalled in the next two sections of this chapter in order to equip the reader with the notions needed for the remainder of the book. The next two sections can be safely skipped by readers already acquainted with ToF technology and active triangulation. The depth or distance measurements taken by the systems of Figure 1.1 can be typically converted into depth maps, i.e., data with each spatial coordinate (x, y) associated to depth information z, and the depth maps can be combined into full all-around 3D models [2].

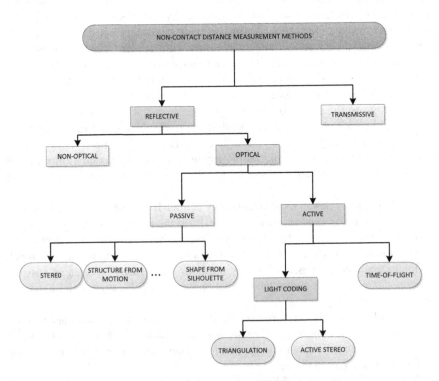

Fig. 1.1 Taxonomy of distance measurement methods (derived from [1]).

1.1 Basics of ToF sensors

A point-wise ToF sensor estimates its *radial* distance from a scene point by the Time-of-Flight or RADAR (Radio Detection And Ranging) principle. In simple words, since the electro-magnetic radiation travels in the air at light speed $c \approx 3 \times 10^8 \, [m/s]$, the distance ρ covered at time τ by an optical radiation is $\rho = c\tau$. Figure 1.2 shows the typical ToF measurement scheme: the radiation emitted at time 0 by the ToF sensor transmitter on the left travels straight towards the scene for a distance ρ, it is then reflected back by the scene surface, it travels back again for a distance ρ and at time τ it reaches the ToF sensor receiver, ideally co-positioned with the transmitter. Since at time τ the path length covered by the radiation is 2ρ,

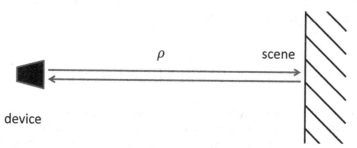

Fig. 1.2 ToF operating principle.

the relationship between ρ and τ in this case is

$$\rho = \frac{c\tau}{2} \tag{1.1}$$

which is the basis of ToF cameras distance measurements.

In order to measure scene surfaces rather than scene points, a number of distance measurement systems mount a point-wise ToF sensor on a scene scanning mechanism. It is rather typical moving the ToF sensor along a linear support, as in the case of air-bone land surveillance LIDARs (Light Detection And Ranging) systems, or along a vertical linear support placed on a rotating platform, with consequent motion both in vertical and horizontal directions, as in the case of the scan-systems used for topographic or architectural surveys (e.g., [3, 4, 5, 6]). Since any time-sequential scanning mechanism takes some time in order to scan a scene, such systems are intrinsically unsuited to acquire dynamic scenes, i.e., scenes with moving objects. Differently from the systems which acquire the scene geometry by a point-wise ToF sensor mounted on time-sequential scanning mechanisms, matricial ToF cameras estimate the scene geometry in a single shot by a matrix of $N_R \times N_C$ ToF sensors (where N_R is the number of matrix rows and N_C the number of columns) each one independently but simultaneously measuring the distance of a scene point in front of them. ToF cameras deliver depth maps at video rates, or measurement matrices with entries giving the distance between the matrix pixel and the corresponding scene point.

In spite of the conceptual simplicity of relationship (1.1), its implementation hides tremendous technological challenges because it involves the light speed. For example, since according to (1.1) it takes $5[ps]$ to cover a $1[mm]$ path, distance measurements of nominal distance resolution of $1[mm]$ need a *clock* capable to measure $5[ps]$ time steps. The implementation of this type of devices and their integration into matricial configurations are fundamental issues of current ToF systems development. Different *clock technology* choices lead to different ToF cameras types. The most common choices are the continuous wave (CW) intensity modulation approach introduced in [7], the optical shutter (OS) approach of [8, 9] and the single-photon avalanche diodes (SPAD) approach [10]. Since all the commercial solutions of [11, 12, 13] are based on CW-ToF technology, this book focuses only on this technology, which is presented in detail in Chapter 2. An exhaustive review of the state-of-the-art in ToF technology can be found in [14].

1.2 Basics of imaging systems and Kinect^{TM} operation

The Kinect^{TM} is a special case of 3D acquisition systems based on light coding. Understanding its operation requires a number of preliminary notions, such as the concepts of *pin-hole camera model*, *camera projection matrix* and *triangulation*, which are given in the next section. The technology behind the Kinect^{TM} will then be described in detail in Chapter 3.

1.2.1 Pin-hole camera model

Consider a 3D reference system (with axes x, y and z), called *Camera Coordinates System (CCS)*, with origin at O, called *center of projection*, and a plane parallel to the (x,y)-plane intersecting the z-axis at negative z-coordinate f, called *sensor* or *image plane S* as shown in Figure 1.3. The axis orientations follow the so-called right-hand convention. Consider also a 2D reference system

$$u = x + c_x \tag{1.2}$$

$$v = y + c_y \tag{1.3}$$

associated to the sensor, called *S-2D reference system*, oriented as shown in Figure 1.3 a. The intersection c of the z-axis with the sensor plane has coordinates $\mathbf{c} = [u = c_x, v = c_y]^T$. The set of sensor points p, called *pixels*, of coordinates $\mathbf{p} = [u,v]^T$ obtained from the intersection of the rays connecting the center of projection O with all the 3D scene points P with coordinates $\mathbf{P} = [x,y,z]^T$ is the scene footprint on the sensor S. The relationship between P and p, called *central* or *perspective projection*, can be readily shown by triangles similarity (see Figure 1.3 b and c) to be

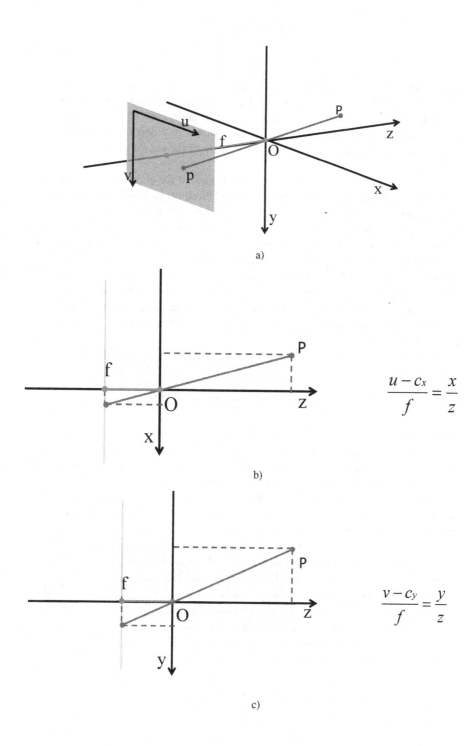

a)

$$\frac{u - c_x}{f} = \frac{x}{z}$$

b)

$$\frac{v - c_y}{f} = \frac{y}{z}$$

c)

Fig. 1.3 Perspective projection: a) scene point P is projected to sensor pixel p; b)horizontal section of a); c) vertical section of a).

$$
\begin{cases}
u - c_x = f\frac{x}{z} \\
\\
v - c_y = f\frac{y}{z}
\end{cases}
\tag{1.4}
$$

where the distance $|f|$ between the sensor plane and the center of projection O is typically called *focal length*. In the adopted notation f is the negative coordinate of the location of the sensor plane with respect to the z-axis. The reader should be aware that other books adopt a different notation, where f denotes the focal length, hence it is a positive number and the z coordinate of the sensor plane is denoted as $-f$.

Perspective projection (1.4) is a good description of the geometrical relationship between the coordinates of the scene points and those of an image of them obtained by a pin-hole imaging device with pin-hole positioned at center of projection O. Such a system allows a single light ray to go throughout the pin-hole at O. For a number of reasons in imaging systems it is more practical to use optics, i.e., suitable sets of lenses, instead of pin-holes. Quite remarkably the ideal model of an optics, called *thin-lens model*, maintains relationship (1.4) between the coordinates of P and of p if the lens optical center (or *nodal point*) is in O and the lens optical axis, i.e., the line orthogonally intersecting the lens at its nodal point, is orthogonal to the sensor. If a thin lens replaces a pin-hole in Fig. 1.3 c, the optical axis coincides with the z-axis of the CCS.

1.2.2 Intrinsic and extrinsic camera parameters

Projective geometry associates to each 2D point p with Cartesian coordinates $\mathbf{p} = [u,v]^T$ of a plane a 3D representation, called *homogeneous coordinates* $\tilde{\mathbf{p}} = [hu, hv, h]^T$, where h is any real constant. The usage of $h = 1$ is rather common and $[u,v,1]^T$ is often called the *extended vector of p* [15].
The coordinates of $\mathbf{p} = [u,v]^T$ can be obtained from those of $\tilde{\mathbf{p}} = [hu, hv, h]^T$ dividing them by the third coordinate h. Vector $\tilde{\mathbf{p}}$ can be interpreted as the 3D ray connecting the sensor point p with the center of projection O.
In a similar way each 3D point P with Cartesian coordinates $\mathbf{P} = [x,y,z]^T$ can be represented in homogeneous coordinates by a 4D vector $\tilde{\mathbf{P}} = [hx, hy, hz, h]^T$ where h is any real constant. Vector $[x,y,z,1]^T$ is often called the *extended vector of P*.
The coordinates of $\mathbf{P} = [x,y,z]^T$ can be obtained from $P = [hx, hy, hz, h]^T$ dividing them by the fourth coordinate h. An introduction to projective geometry suitable to its computer vision applications can be found in [16].

The homogeneous coordinates representation of p allows to rewrite non-linear relationship (1.4) in a convenient matrix form, namely:

$$
z \begin{bmatrix} u \\ v \\ 1 \end{bmatrix} = \begin{bmatrix} f & 0 & c_x \\ 0 & f & c_y \\ 0 & 0 & 1 \end{bmatrix} \begin{bmatrix} x \\ y \\ z \end{bmatrix}
\tag{1.5}
$$

Note that the left side of (1.5) represents p in 2D homogeneous coordinates but the right side of (1.5) represents P in 3D cartesian coordinates. It is straightforward to add a column with all 0 entries at the right of the matrix in order to represent P in homogeneous coordinates too. This latter representation is more common than (1.5), which nevertheless is often adopted [15] for its simplicity.

Digital sensor devices are typically planar matrices of rectangular sensor cells hosting photoelectric conversion systems in CMOS or CCD technology in the case of photo-cameras or video-cameras or single ToF receivers in the case of ToF cameras, as explained in the next chapter. Customarily they are modeled as a rectangular lattice Λ_S with horizontal and vertical step-size k_u and k_v respectively as shown in Figure 1.4 a.

Given the finite sensor size, only a rectangular window of Λ_S made by N_C columns and N_R rows is of interest for imaging purposes. In order to deal with normalized

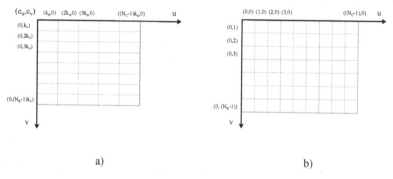

a)   b)

Fig. 1.4 2D sensor coordinates: a) rectangular window of a non-normalized orthogonal lattice; b) rectangular window of a normalized orthogonal lattice.

lattices with origin at $(0,0)$ and unitary pixel coordinates $\mathbf{u_S} \in [0,...,N_C-1]$ and $\mathbf{v_S} \in [0,...,N_R-1]$ in both u and v direction, relationship (1.5) is replaced by

$$z \begin{bmatrix} u \\ v \\ 1 \end{bmatrix} = K \begin{bmatrix} x \\ y \\ z \end{bmatrix} \tag{1.6}$$

where K is the intrinsic parameters matrix defined as

$$K = \begin{bmatrix} f_x & 0 & c_x \\ 0 & f_y & c_y \\ 0 & 0 & 1 \end{bmatrix} \tag{1.7}$$

with $f_x = fk_u$ the x-axis focal length of the optics, $f_y = fk_v$ the y-axis focal length of the optics, c_x and c_y the (u,v) coordinates of the intersection of the optical axis with the sensor plane. All these quantities are expressed in $[pixel]$, i.e., since f is in $[mm]$, k_u and k_v are assumed to be $[pixel]/[mm]$.

In many practical situations it is convenient to represent the 3D scene points not with respect to the CCS, but with respect to a different easily accessible reference system conventionally called *World Coordinate System (WCS)*, in which a scene point denoted as P_W has coordinates $\mathbf{P_W} = [x_W, y_W, z_W]^T$. The relationship between the representation of a scene point with respect to the CCS, denoted as P, and its representation with respect to the WCS, denoted as P_W is

$$\mathbf{P} = R\mathbf{P_W} + \mathbf{t} \tag{1.8}$$

where R and $\mathbf{t}$ are a suitable rotation matrix and translation vector respectively. By representing P_W at the right side in homogeneous coordinates $\tilde{\mathbf{P}}_\mathbf{W} = [hx_W, hy_W, hz_W, h]^T$ and choosing $h = 1$, relationship (1.8) can be rewritten as

$$\mathbf{P} = [R\ \mathbf{t}]\tilde{\mathbf{P}}_\mathbf{W} \tag{1.9}$$

In this case the relationship between a scene point represented in homogeneous coordinates with respect to the WCS and its corresponding pixel in homogeneous coordinates too, from (1.6) becomes

$$z\begin{bmatrix} u \\ v \\ 1 \end{bmatrix} = K\mathbf{P} = K[R\ \mathbf{t}]\tilde{\mathbf{P}}_\mathbf{W} = M\tilde{\mathbf{P}}_\mathbf{W} = M\begin{bmatrix} x_W \\ y_W \\ z_W \\ 1 \end{bmatrix} \tag{1.10}$$

where the 3×4 matrix

$$M = K[R\ \mathbf{t}] \tag{1.11}$$

is called *projection matrix*. Matrix M depends on the intrinsic parameters matrix K and on the extrinsic parameters R and $\mathbf{t}$ of the imaging system.

As a consequence of distortions and aberrations of real optics, the coordinates $\hat{\mathbf{p}} = (\hat{u}, \hat{v})$ of the pixel actually associated to scene point P with coordinates $\mathbf{P} = [x, y, z]^T$ in the CCS system do not satisfy relationship (1.6). The correct pixel coordinates (u, v) of (1.6) can be obtained from the distorted coordinates $(\hat{u}, \hat{v})$ actually measured by the imaging system by inverting suitable distortion models, i.e., as $\mathbf{p_T} = \Psi^{-1}(\hat{\mathbf{p}}_\mathbf{T})$, where $\Psi(\cdot)$ denotes the distortion transformation. Anti-distortion model (1.12), also called *Heikkila model*, has become popular since it is adequate for the distortions of most imaging systems and there are effective methods for computing its parameters [17]:

$$\begin{bmatrix} u \\ v \end{bmatrix} = \Psi^{-1}(\hat{\mathbf{p}}_\mathbf{T}) = \begin{bmatrix} \hat{u}(1 + k_1 r^2 + k_2 r^4 + k_3 r^6) + 2d_1\hat{v} + d_2(r^2 + 2\hat{u}^2) \\ \hat{v}(1 + k_1 r^2 + k_2 r^4 + k_3 r^6) + d_1(r^2 + 2\hat{v}^2) + 2d_2\hat{u} \end{bmatrix} \tag{1.12}$$

where $r = \sqrt{(\hat{u} - c_x)^2 + (\hat{v} - c_y)^2}$, parameters k_i with $i = 1, 2, 3$ are constants accounting for radial distortion and d_i with $i = 1, 2$ for tangential distortion. A number of other more complex models, e.g. [18], are also available.
Distortion parameters

$$\mathbf{d} = [k_1, k_2, k_3, d_1, d_2]$$

(1.13)

are intrinsic parameters to be considered together with $[f, k_u, k_v, c_x, c_y]$.

The estimation of intrinsic and extrinsic parameters of an imaging system is called *geometrical calibration* as discussed in Chapter 4. Suitable tools for this task are given by [19] and [20].

1.2.3 Stereo vision systems

A stereo vision (or just *stereo*) system is made by two standard (typically identical) cameras partially framing the same scene. Such a system can always be calibrated and rectified [15]. It then becomes equivalent to a stereo vision system made by two (identical) standard cameras with coplanar and aligned imaging sensors and parallel optical axis. Chapter 4 reports about popular calibration and rectification procedures. The theory behind them, known as *epipolar geometry*, can be found in classical computer vision books such as [15, 16].

Let us introduce next the minimal stereo vision notation. The two cameras of the (calibrated and rectified) stereo vision system S are the left camera L (also called *reference camera*) and the right camera R (also called *target camera*). Each camera, as seen above, has its own 3D CCS and 2D reference systems as shown in Figure 1.5. Namely the L camera has CCS with coordinates (x_L, y_L, z_L), also called *L-3D reference system*, and a 2D reference system with coordinates (u_L, v_L). The R camera has CCS with coordinates (x_R, y_R, z_R), also called *R-3D reference system*, and a 2D reference system with coordinates (u_R, v_R). A common convention is to consider

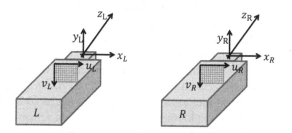

Fig. 1.5 Stereo vision system coordinates and reference systems.

the L-3D reference system as the reference system of the stereo vision system, and to denote it as S-3D reference system.

In the case of a calibrated and rectified stereo vision system, a 3D point P with coordinates $\mathbf{P} = [x, y, z]^T$ with respect to the S-3D reference system is projected to the pixels p_L and p_R of the L and R cameras with coordinates $\mathbf{p_L} = [u_L, v_L]^T$ and $\mathbf{p_R} = [u_R = u_L - d, v_R = v_L]^T$ respectively as shown in Figure 1.6. From the relationships concerning the triangle with vertices at p_R, P and p_L it can be shown that in a rectified setup, where points p_L and p_R have the same vertical coordinates, the difference between their horizontal coordinates $d = u_L - u_R$, called *disparity*, is inversely proportional to the depth value z of P through the well-known triangulation relationship

$$z = \frac{b|f|}{d} \qquad (1.14)$$

The reader interested in the derivation of Equation 1.14 is referred to [15, 16]. In (1.14) b is the *baseline*, i.e, the distance between the origins of the L-3D and the R-3D reference systems, and $|f|$ is the focal length of both cameras. Pixels p_L and p_R are called *conjugate*. From the 2D coordinates of p_L and the associated depth z obtained from (1.14), the coordinates x and y of the corresponding 3D point P represented with respect to the CCS can be computed by inverting projection equation (1.6) relatively to camera L, i.e.:

$$\begin{bmatrix} x \\ y \\ z \end{bmatrix} = K_L^{-1} \begin{bmatrix} u_L \\ v_L \\ 1 \end{bmatrix} z \qquad (1.15)$$

where K_L^{-1} is the inverse of the rectified intrinsic parameters matrix (1.7) of camera L. Therefore, once a couple of conjugate pixels p_L and p_R of a stereo image pair becomes available, from them the 3D coordinates $\mathbf{P} = [x, y, z]^T$ of a scene point P can be computed from (1.14) and (1.15), usually called *triangulation* or *computational stereopsis*.

The availability of a pair of conjugate pixels is a tricky part of the above procedure, first of all because such a pair may not exist because of occlusions and even if it exists it may not be straightforward finding it.

Detecting conjugate pixels between the stereo image pair, typically called the *correspondence problem*, is one of the major challenges of a stereo vision algorithm. The methods proposed for this task are typically divided in *local* and *global* approaches. Local methods consider only local similarity measures between the region surrounding p_L and regions of similar shape around all the candidate conjugate points p_R of the same row. The selected conjugate point is the one maximizing the similarity measure, a method typically called *Winner Takes All (WTA)* strategy.

Global methods do not consider each couple of points on its own but estimate all the disparity values at once exploiting global optimization schemes. Global methods based on Bayesian formulations are currently receiving great attention. Such techniques generally model the scene as a Markov random field (MRF) and include within a unique framework cues coming from the local comparisons between the two images and scene depth smoothness constraints. Global stereo vision algorithms typically estimate the disparity image by minimizing a cost function made by a *data term* representing the cost of local matches, similar to the one of local algorithms (e.g., covariance) and a *smoothness term* defining the smoothness level of the disparity image by explicitly or implicitly accounting for discontinuities [15].

Although specific algorithms may have a considerable impact on the solution of the correspondence problem, the ultimate quality of 3D stereo reconstruction

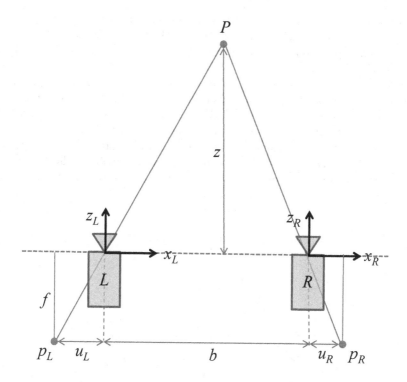

Fig. 1.6 Triangulation with a pair of aligned and rectified cameras.

inevitably depends also on scene characteristics. This can be readily realized considering the case of a scene without geometric or color features, such as a straight wall of uniform color. The stereo images of such a scene will be uniform, and since no corresponding pixels can be obtained from them, no depth information about the scene can be obtained by triangulation. Active triangulation used in the so called *light coding systems* introduced next, offers an effective way to cope with the correspondence problem issues.

1.2.4 Light coding systems

Equation (1.14) derives from the relationships concerning the triangles in Figure 1.6. The fact that p_L and p_R in standard stereo systems are due to the light reflected by P towards the two cameras is secondary. The main point is the triangle geometry between rays Pp_L, Pp_R and $p_L p_R$. In this perspective inspired by a projective geometry point of view, in which image points are equivalent to rays exiting two centers of projection, any device capable of projecting rays between its center of projection and the scene points can be considered functionally equivalent to a standard camera. This is the case of light projectors, since any light ray emitted by them connects their center of projection to the scene point P by the light pattern pixel p_A projected to P, as shown in Figure 1.7.

A stereo vision system where one of the two cameras is replaced by a projector, i.e., made by a camera C and a projector A as shown in Figure 1.7, is called *active* or *light coding system*. Camera C has CCS system with coordinates (x_C, y_C, z_C) also called *C-3D reference system* and a 2D reference system with coordinates (u_C, v_C) as shown in Figure 1.7. Projector A similarly has CCS with coordinates (x_A, y_A, z_A) also called *A-3D reference system* and a 2D reference system with coordinates (u_A, v_A). As in the case of standard passive stereo systems, active systems of the type shown

$$\mathbf{P} = [\mathrm{x, y, z}]^{\mathrm{T}}$$

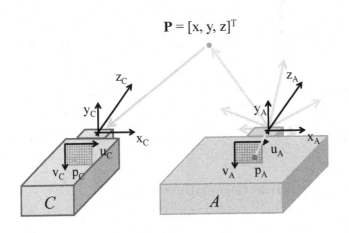

Fig. 1.7 Active triangulation by a system made by a camera C (blue) and a light projector A (green).

in Figure 1.7 can be calibrated and rectified [21] in order to simplify the depth estimation process.

Figure 1.7 shows the projection of light pattern pixel p_A with coordinates $\mathbf{p_A} = [u_A, v_A]^T$ in the A-2D reference system to 3D scene point P with coordinates $\mathbf{P} = [x, y, z]^T$ in the C-3D reference system. If P is not occluded it projects the light radiant power received by the projector to pixel p_C of camera C establishing triangle $p_C P p_A$. If the active system is calibrated and rectified, p_C has coordinates $\mathbf{p_C} = [u_C = u_A + d, v_C = v_A]$. As in the case of standard stereo systems, since p_A and p_C are conjugate points, once their coordinates are known the depth z of P can be computed from (1.14), which in this case is called *active triangulation* since A is an active system, and the 3D coordinates of P can be computed from (1.15) as above.

The effectiveness of active systems with respect to the correspondence problem can be easily appreciated in the previously considered example of a straight wall of uniform color. In this case the light pattern pixel p_A of the projector "colors" with its radiant power the scene point P to which it projects. In this way the pixel p_C of the camera C where P is projected (obviously in presence of a straight wall there are not occlusions) receives from P the "color" of p_A and becomes recognizable among its neighboring pixels. In this case conjugate points p_A and p_C exist and can also be easily detected by adopting suitable light patterns [21].

Light coding systems can therefore provide depth information also in case of scenes without geometry and color features where standard stereo systems fail to give any depth data.

The characteristics of the projected patterns are fundamental for the correspondence problem solution and the overall system performance and their study attracted a lot of attention. The projection of sequences of various tens of light patterns was typical of early light coding methods and this limited the range of measurable distances and their usage to still scenes. This type of systems were and continue to be the common choice for 3D modeling of still scenes. For the 3D modeling methods used in this field see for instance [2, 22]. In general, active techniques are recognized to be more expensive and slower than passive methods but way more accurate and robust than them. In order to measure distances of dynamic scenes, i.e., scenes with moving objects, subsequent light coding methods focused on reducing the number of projected patterns to few units or to a single pattern, e.g., [23, 24, 25]. The Kinect^{TM} belongs to this last family of methods as seen in greater detail in Chapter 3.

1.3 Plan of the book

This introduction motivates the book and provides the basics for understanding the ToF sensors and the imaging systems behind light coding systems and Kinect^{TM}.

The two subsequent chapters are devoted to the operation principles of ToF cameras and Kinect^{TM}, both because they are not yet widely known and also because this helps understanding the general characteristics of their data (e.g., the high noise and limited resolution typical of current ToF camera data or the strong edge-artifacts of current Kinect^{TM} data).

Chapter 2 describes the continuous wave ToF sensors operation, their practical issues and the projection aspects of ToF cameras, i.e., of the imaging systems supporting such sensors. Chapter 3 considers the operation of Kinect^{TM} and the artifacts typical of its depth measurements.

Given the relatively short life-time of matricial ToF cameras and Kinect^{TM} technologies it cannot come as a surprise that the current quality of their depth data may not be comparable with that of the images provided by today's photo-cameras or video-cameras (simply called *standard cameras* in the sequel). In front of the high quality/price ratio offered by image and video technology, the idea of combining low-resolution and high noise depth data with high resolution and low-noise images is rather intriguing. The interest is both of the conceptual nature for its methodological implications and of the practical nature for the substantial depth data improvements at very low costs this synergy promises. The second part of this book focuses on the data processing methods suited to depth information and to the combination of this information with standard images and video streams.

Clearly the effective joint usage of depth and color data streams requires first the calibration of all the deployed acquisition devices and their accurate registration. This is a fundamental and delicate issue, addressed in Chapter 4. Chapter 4 considers first the calibration of the single devices (standard cameras, ToF cameras, Kinect) and then the joint calibration between a ToF camera or Kinect^{TM} and one or more standard cameras.

Chapter 5 presents methods for improving the characteristics of an original depth data stream, such as spatial resolution, depth resolution accuracy, signal-to-noise ratio, edge-artifacts and similar with the assistance of one or two standard cameras. In particular improving the spatial resolution of depth data (operation typically referred as *super-resolution*) is of special interest since one of the major drawbacks of the data currently produced by ToF cameras is their low spatial resolution. Super-resolution can be obtained just by pairing a depth camera with a single standard camera, which is a setup extremely interesting for its simplicity and inexpensiveness. The super-resolution processing techniques used for depth data often extend methods originally proposed for images by suitably taking into account, together with the high resolution color stream, the depth nature of the considered data, which is rather different from the radiometric intensity nature of images. Data fusion refers to the more general task of synergycally combining two or more depth descriptions of the same scene captured by different systems operating synchronously but independently with the purpose of delivering a unique output stream with characteristics (e.g., resolution accuracy, signal-to-noise ratio, edge-artifacts and similar) improved with respect to those of the original inputs. The data fusion procedures introduced in Chapter 5 concern setups made by a depth camera and a pair of standard cameras, i.e., concern a depth data stream coming from a depth camera and another one coming from a stereo vision algorithm fed by the two standard cameras. A major point of interest of this approach is the vast variety of stereo vision techniques available for this application, each one contributing different cost/performance characteristics.

The last chapter of this book presents an example of application where depth data can be a major asset: scene segmentation, or the recognition of the regions corresponding to the different scene objects. This is a classical problem traditionally dealt by way of images. Unfortunately this approach, called *image segmentation*, is an ill-posed problem, not completely solved after decades of research. The use of depth together with color can drastically simplify the segmentation task and deliver segmentation tools based on combined depth and color information which outperform the segmentation techniques based on a single cue only (either depth or color). This fact has received special attention in *video-matting*, a strictly related application (namely the separation of foreground objects from the background) widely used in the film-making industry.

References

1. J.-Y. Bouguet, B. Curless, P. Debevec, M. Levoy, S. Nayar, and S. Seitz, "Overview of active vision techniques. siggraph 2000 course on 3d photography." Workshop, 2000.
2. F. Bernardini and H. Rushmeier, "The 3d model acquisition pipeline," *Comput. Graph. Forum*, vol. 21, no. 2, pp. 149–172, 2002.
3. "Leica." http://hds.leica-geosystems.com.
4. "Riegl." http://www.riegl.com/.
5. "Faro." http://faro.com.
6. "Zoller and Frolich." http://www.zf-laser.com/.
7. R. Lange, *3D Time-Of-Flight distance measurement with custom solid-state image sensors in CMOS/CCD-technology.* PhD thesis, University of Siegen, 2000.
8. G. Iddan and G. Yahav, "G.: 3d imaging in the studio (and elsewhere," *In: SPIE*, pp. 48–55, 2001.
9. G. Yahav, G. Iddan, and D. Mandelboum, "3d imaging camera for gaming application," in *ICCE*, 2007.
10. L. Pancheri, N. Massari, F. Borghetti, and D. Stoppa, "A 32x32 spad pixel array with nanosecond gating and analog readout," in *International Image Sensor Workshop (IISW)*, 2011.
11. "Mesa imaging." http://www.mesa-imaging.ch.
12. "Pmd technologies." http://www.pmdtec.com/.
13. "Softkinetic." http://www.softkinetic.com/.
14. D. Stoppa and F. Remondino, eds., *TOF Range-Imaging Cameras.* Springer, 2012.
15. R. Szeliski, *Computer Vision: Algorithms and Applications.* New York: Springer, 2010.
16. R. Hartley and A. Zisserman, *Multiple View Geometry in Computer Vision.* Cambridge University Press, 2004.
17. J. Heikkila and O. Silven, "A four-step camera calibration procedure with implicit image correction," in *CVPR*, 1997.
18. D. Claus and A. Fitzgibbon, "A rational function lens distortion model for general cameras," in *CVPR*, 2005.
19. J.-Y. Bouguet, "Camera calibration toolbox for matlab."
20. "OpenCV." http://opencv.willowgarage.com/wiki/.
21. M. Trobina, "Error model of a coded-light range sensor," tech. rep., Communication Technology Laboratory Image Science Group, ETH-Zentrum, Zurich, 1995.
22. B. Curless and M. Levoy, "A volumetric method for building complex models from range images," in *Proceedings of the 23rd annual conference on Computer graphics and interactive techniques*, SIGGRAPH '96, (New York, NY, USA), pp. 303–312, ACM, 1996.
23. L. Zhang, B. Curless, and S. Seitz, "Spacetime stereo: shape recovery for dynamic scenes," in *Proceedings. 2003 IEEE Computer Society Conference on Computer Vision and Pattern Recognition, CVPR 2003.*, 2003.
24. L. Zhang, B. Curless, and S. Seitz, "Rapid shape acquisition using color structured light and multi-pass dynamic programming," in *In The 1st IEEE International Symposium on 3D Data Processing, Visualization, and Transmission*, pp. 24–36, 2002.
25. J. Salvi, J. Pags, and J. Batlle, "Pattern codification strategies in structured light systems," *Pattern Recognition*, vol. 37, pp. 827–849, 2004.

Chapter 2
CW Matricial Time-of-Flight Range Cameras

Matricial Time-of-Flight range cameras (simply *ToF cameras* in this book) are active sensors capable to acquire the three-dimensional geometry of the framed scene at video rate (up to $50\,[fps]$). Commercial products are currently available from independent manufacturers, such as MESA Imaging [1] (Figure 2.1), PMD Technologies [2] and Optrima SoftKinetic [3]. Microsoft [4] is another major actor in the ToF camera technology arena since at the end of 2010 it acquired Canesta, a U.S. ToF camera manufacturer. Other companies (e.g., Panasonic [5] and IEE [6]) and research institutions (e.g., CSEM [7] and Fondazione Bruno Kessler [8]) are also working on ToF cameras.

Fig. 2.1 Example of commercial ToF camera: MESA Imaging SR4000^{TM}.

As anticipated in Section 1.1, this chapter examines the continuous wave ToF technology, the one adopted in the sensors of all the current commercial products. Section 2.1 presents the operating principles of such technology and Section 2.2 the practical issues at the basis of its performance limits and noise characteristics. The characteristics of ToF cameras, i.e., of the imaging system supporting ToF sensors, are considered in Section 2.3. Section 2.4 has the conclusion and the further reading.

2.1 CW ToF sensors: operation principles

According to the scheme of Figure 1.2, continuous wave ToF cameras send towards the scene an infra-red (IR) optical signal $s_E(t)$ of amplitude A_E modulated by a sinusoid of frequency f_{mod}, namely

$$s_E(t) = A_E[1 + \sin(2\pi f_{mod}t)] \tag{2.1}$$

Signal $s_E(t)$ is reflected back by the scene surface and travels back towards a receiver co-positioned with the emitter.

The signal reaching the receiver, because of the energy absorption generally associated to the reflection, because of free-path propagation attenuation (proportional to the square of the distance) and because of the non-instantaneous propagation of IR optical signals leading to a phase delay $\Delta\phi$, can be written as

$$s_R(t) = A_R[1 + \sin(2\pi f_{mod}t + \Delta\phi)] + B_R \tag{2.2}$$

where A_R is the attenuated amplitude of the received signal and B_R is the interfering radiation at the IR wavelength of the emitted signal reaching the receiver. Figure 2.2 shows an example of emitted and received signals. Quantity A_R (from now denoted

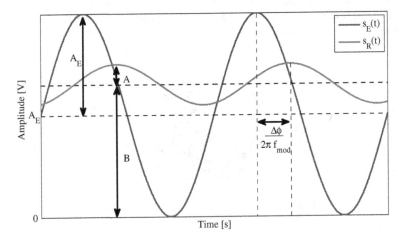

Fig. 2.2 Example of emitted signal $s_E(t)$ (in blue) and received signal $s_R(t)$ (in red).

by A) is called *amplitude*, since it is the amplitude of the useful signal. Quantity $A_R + B_R$ (from now denoted by B) is called *intensity* or *offset*, and it is the average[1] of the received signal (with a component A_R due to the modulation carrier and

---

[1] It is common to call A and B *amplitude* and *intensity* respectively, even though both A and B are IR radiation amplitudes (measured in $[V]$). A is also the amplitude of the received sinusoidal signal.

an interference component B_R due to background illumination). According to this notation, Equation (2.2) can be rewritten as

$$s_R(t) = A\sin(2\pi f_{mod}t + \Delta\phi) + B \qquad (2.3)$$

The unknowns of Equation (2.3) are A, B and $\Delta\phi$, where A and B as IR radiation amplitudes are measured in volt $[V]$ and $\Delta\phi$ as a phase value is a pure number. The most important unknown is $\Delta\phi$, since CW ToF cameras infer distance ρ from $\Delta\phi$ from (1.1) and (2.2)

$$\Delta\phi = 2\pi f_{mod}\tau = 2\pi f_{mod}\frac{2\rho}{c} \qquad (2.4)$$

or equivalently

$$\rho = \frac{c}{4\pi f_{mod}}\Delta\phi \qquad (2.5)$$

Unknowns A and B as it will be seen are important for SNR considerations.

In order to estimate the unknowns A, B and $\Delta\phi$, the receiver samples $s_R(t)$ at least 4 times per period of the modulating signal [9]. For instance, if the modulation frequency is $30[MHz]$, the received signal must be sampled at least at $120[MHz]$. Assuming a sampling frequency $F_S = 4f_{mod}$, given the 4 samples per period $s_R^0 = s_R(t = 0)$, $s_R^1 = s_R(t = 1/F_S)$, $s_R^2 = s_R(t = 2/F_S)$ and $s_R^3 = s_R(t = 3/F_S)$, the receiver estimates values $\hat{A}$, $\hat{B}$ and $\widehat{\Delta\phi}$ as

$$(\hat{A}, \hat{B}, \widehat{\Delta\phi}) = \arg\min_{A,B,\Delta\phi} \sum_{n=0}^{3}\{s_R^n - [A\sin(\frac{\pi}{2}n + \Delta\phi) + B]\}^2 \qquad (2.6)$$

As described in [10] and [11], after some algebraic manipulations from (2.6) one obtains

$$\hat{A} = \frac{\sqrt{(s_R^0 - s_R^2)^2 + (s_R^1 - s_R^3)^2}}{2} \qquad (2.7)$$

$$\hat{B} = \frac{s_R^0 + s_R^1 + s_R^2 + s_R^3}{4} \qquad (2.8)$$

$$\widehat{\Delta\phi} = \arctan2\left(s_R^0 - s_R^2, s_R^1 - s_R^3\right) \qquad (2.9)$$

The final distance estimate $\hat{\rho}$ can be obtained combining (2.5) and (2.9) as

$$\hat{\rho} = \frac{c}{4\pi f_{mod}}\widehat{\Delta\phi} \qquad (2.10)$$

2.2 CW ToF sensors: practical implementation issues

The above derivation highlights the conceptual steps needed to measure the distance ρ of a scene point from a CW ToF sensor, with co-positioned emitter and receiver.

In practice a number of non-idealities, such as *phase wrapping, harmonic distortion, noise sources, saturation* and *motion blur*, must be taken into account.

2.2.1 Phase wrapping

The first fundamental limitation of CW ToF sensors comes from the fact that the estimate of $\widehat{\Delta\phi}$ is obtained from an arctangent function, which as well-known has codomain $[-\frac{\pi}{2}, \frac{\pi}{2}]$. Therefore the estimates of $\widehat{\Delta\phi}$ can only assume values in this interval. Since the physical delays entering the phase shift $\Delta\phi$ of Equation (2.4) can only be positive, it is possible to shift the $\arctan(\cdot)$ codomain to $[0, \pi]$ in order to have a larger interval available for $\widehat{\Delta\phi}$. Moreover, the usage of $\arctan2(\cdot, \cdot)$ allows to extend the codomain to $[0, 2\pi]$. From Equation (2.10) it is immediate to see that the estimated distances are within range $[0, \frac{c}{2f_{mod}}]$. If for instance $f_{mod} = 30[MHz]$, the interval of measurable distances is $[0 - 5][m]$.

Since $\widehat{\Delta\phi}$ is estimated modulo 2π from (2.10) and the distances greater than $\frac{c}{2f_{mod}}$ correspond to $\widehat{\Delta\phi}$ greater than 2π, they are wrongly estimated. In practice the distance returned by (2.10) corresponds to the remainder of the division between the actual $\Delta\phi$ and 2π, multiplied by $\frac{c}{2f_{mod}}$, a well-known phenomenon called *phase wrapping* since it may be ragarded as a periodic wrapping around 2π of phase values $\widehat{\Delta\phi}$. Clearly if f_{mod} increases, the interval of measurable distances becomes smaller, and vice-versa. Possible solutions to overcome phase wrapping include the usage of multiple modulation frequencies or of non-sinusoidal wave-forms (e.g., chirp wave-forms).

2.2.2 Harmonic distortion

The generation of perfect sinusoids of the needed frequency is not straightforward. In practice [12], actual sinusoids are obtained as low-pass filtered versions of squared wave-forms emitted by LEDs. Moreover, the sampling of the received signal is not ideal, but it takes finite time intervals, as shown in Figure 2.3. The combination of these two factors introduces an harmonic distortion in the estimated phase-shift $\widehat{\Delta\phi}$ and consequently in the estimated distance $\hat{\rho}$. Such harmonic distortion leads to a systematic offset component dependent on the measured distance. A metrological characterization of this harmonic distortion effect is reported in [13] and [14].

Figure 2.4 shows that the harmonic distortion offset exhibits a kind of oscillatory behavior which can be up to some tens of centimeters, clearly reducing the accuracy of distance measurements. As reported in Chapter 4, this systematic offset can be fixed by a look-up-table (LUT) correction.

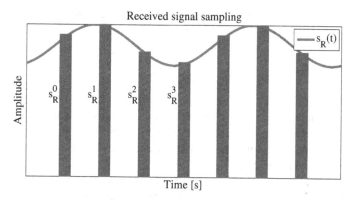

Fig. 2.3 Pictorial illustration of non instantaneous sampling of the received signal $s_R(t)$.

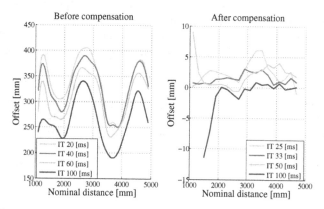

Fig. 2.4 Left: systematic distance measurements offset due to harmonic distortion before compensation (from [13]). Right: systematic distance measurements offset after compensation (courtesy of MESA Imaging).

2.2.3 Photon-shot noise

Because of the light-collecting nature of the receiver, the acquired samples s_R^0, s_R^1, s_R^2 and s_R^3 are affected by photon-shot noise, due to dark electron current and photon-generated electron current as reported in [10]. Dark electron current can be reduced by lowering the sensor temperature or by technological improvements. Photon-generated electron current, due to light-collection, cannot be completely eliminated. Photon-shot noise is statistically characterized by a Poisson distribution. Since $\hat{A}$, $\hat{B}$, $\widehat{\Delta\phi}$ and $\hat{\rho}$ are computed directly from the corrupted samples s_R^0, s_R^1, s_R^2 and s_R^3, their noise distribution can be computed by propagating the Poisson distribution through Equations (2.7-2.10). A detailed analysis of error and noise propagations can be found in [11].

Quite remarkably the probability density function of the noise affecting estimate

$\hat{\rho}$ according to [10] and [11] can be approximated by a Gaussian [2] with standard deviation

$$\sigma_\rho = \frac{c}{4\pi f_{mod} \sqrt{2}} \frac{\sqrt{B}}{A} \qquad (2.11)$$

Standard deviation (2.11) determines the precision (repeatability) of the distance measurement and it is directly related to f_{mod}, A and B. In particular, if the received signal amplitude A increases, the precision improves. This suggests that the precision improves as the measured distance decreases and the reflectivity of the measured scene point increases.

Equation (2.11) indicates also that as the interference intensity B of the received signal increases, the precision gets worse. This means that the precision improves as the scene background IR illumination decreases. Note that B may increase because of two factors: an increment of the received signal amplitude A or an increment of the background illumination. While in the second case the precision gets worse, in the first case there is an overall precision improvement, given the squared root dependence of B in (2.11). Finally observe that B cannot be 0 as it depends on carrier intensity A.

If modulation frequency f_{mod} increases the precision improves. The modulation frequency is an important parameter for ToF sensors, since f_{mod} is also related to phase wrapping and to the maximum measurable distance. In fact, if f_{mod} increases the measurement precision improves, while the maximum measurable distance decreases (and vice-versa). Therefore there is a trade-off between distance precision and range. Since generally f_{mod} is a tunable parameter, it can be adapted to the distance precision and range requirements of the specific application.

2.2.4 Other noise sources

There are several other noise sources affecting the distance measurements of ToF sensors, namely *flicker* and a *kTC noise*. The receiver amplifier introduces a Gaussian-distributed thermal noise component. Since the amplified signal is quantized in order to be digitally treated, quantization introduces another error source, customarily modeled as random noise. Quantization noise can be controlled by the number of used bits and it is typically neglectable with respect to the other noise sources. All the noise sources, except photon-shot noise, may be reduced by adopting high quality components. A comprehensive description of the various ToF noise sources can be found in [9, 10, 11, 12].

Averaging distance measurements over several periods is a classical provision to

---

[2] An explicit expression of the Gaussian probability density function mean is not given in [10, 11]. However, the model of [11] provides implicit information about the mean which is a function of both A and B, and contributes to the distance measurement offset. For calibration purposes the non-zero mean effect can be included in the harmonic distortion.

mitigate the noise effects. If N is the number of periods, the estimated values $\hat{A}$, $\hat{B}$ and $\widehat{\Delta\phi}$ become

$$\hat{A} = \frac{\sqrt{\left(\frac{1}{N}\sum_{n=0}^{N-1} s_R^{4n} - \frac{1}{N}\sum_{n=0}^{N-1} s_R^{4n+2}\right)^2 + \left(\frac{1}{N}\sum_{n=0}^{N-1} s_R^{4n+1} - \frac{1}{N}\sum_{n=0}^{N-1} s_R^{4n+3}\right)^2}}{2} \quad (2.12)$$

$$\hat{B} = \frac{\sum_{n=0}^{N-1} s_R^{4n} + \sum_{n=0}^{N-1} s_R^{4n+1} + \sum_{n=0}^{N-1} s_R^{4n+2} + \sum_{n=0}^{N-1} s_R^{4n+3}}{4N} \quad (2.13)$$

$$\widehat{\Delta\phi} = \arctan 2 \left(\frac{1}{N}\sum_{n=0}^{N-1} s_R^{4n} - \frac{1}{N}\sum_{n=0}^{N-1} s_R^{4n+2}, \frac{1}{N}\sum_{n=0}^{N-1} s_R^{4n+1} - \frac{1}{N}\sum_{n=0}^{N-1} s_R^{4n+3} \right) \quad (2.14)$$

where $s_R^{4n} = s_R(4n/F_S)$, $s_R^{4n+1} = s_R((4n+1)/F_S)$, $s_R^{4n+2} = s_R((4n+2)/F_S)$ and $s_R^{4n+3} = s_R((4n+3)/F_S)$.
This provision reduces but does not completely eliminate the noise effects. The averaging intervals used in practice are typically between $1[ms]$ and $100[ms]$. For instance in case of $f_{mod} = 30MHz$, where the modulating sinusoid period is $33.3 \times 10^{-9}[s]$, the averaging intervals concern a number of modulating sinusoid periods from 3×10^4 to 3×10^6. The averaging interval length is generally called *integration time*, and its proper tuning is very important in ToF measurements. Long integration times lead to good ToF distance measurements repeatability.

2.2.5 Saturation and motion blur

Although rather effective against noise, averaging over multiple periods introduces dangerous side effects, such as *saturation* and *motion blur*. Saturation occurs when the received photons quantity exceeds the maximum quantity that the receiver can collect. This phenomenon is particularly noticeable in presence of external IR illumination (e.g., direct solar illumination) or in case of highly reflective objects (e.g., specular surfaces). The longer the integration time, the higher is the quantity of collected photons and the most likely is the possibility of saturation. Specific solutions have been developed in order to avoid saturation, i.e., in-pixel background light suppression and automatic integration time setting [12, 10].

Motion blur is another important phenomenon accompanying time averaging. It is caused, as in the case of standard cameras, by the fact that the imaged objects may move during integration time. Time intervals of the order of $1 - 100[ms]$ make likely objects movement unless the scene is perfectly still. In case of moving objects, the samples entering Equations (2.12 - 2.14) do not concern a specific scene point at subsequent instants as it should be, but different scene points at subsequent instants and expectedly cause distance measurement artifacts. The longer the integration time, the higher the likelihood of motion blur (but better the distance measurement precision). Integration time is another parameter to set in light of the specific application characteristics, needed for their imaging operation.

2.3 Matricial ToF cameras

Let us recall that the ToF sensors considered so far are single devices made by a single emitter and a co-positioned single receiver. Such an arrangement is only functional to single point distance measurements. The structure of actual ToF cameras is more complex than that of the ideal single ToF sensor cells considered so far, both because of the matrix nature of their ToF sensors and because of the optics needed for their imaging operation.

2.3.1 Matricial ToF sensors

A ToF camera sensor may be conceptually interpreted as a matricial organization of a multitude of single devices, each one made by an emitter and a co-positioned receiver as considered so far. In practice implementations based on a simple juxtaposition of a multitude of the previously considered single-point measurement devices are not feasible. Currently it is not possible to integrate $N_R \times N_C$ emitters and $N_R \times N_C$ receivers in a single chip, especially for high values of N_R and N_C as needed in imaging applications. However, it is not true that each receiver requires a specific co-positioned emitter, instead a single emitter may provide an irradiation that is reflected back by the scene and collected by a multitude of receivers close to each other. Once the receivers are separated from the emitters, the former can be implemented as CCD/CMOS lock-in pixels [9, 10] and integrated in a $N_R \times N_C$ matrix. The lock-in pixels matrix is commonly called *ToF camera sensor* (or simply *sensor*), and for example in the case of the MESA Imaging SR4000 it is made by 176×144 lock-in pixels.

The current matricial ToF sensor IR emitters are common LEDs and they cannot be integrated, but can be positioned in a configuration mimicking the presence of a single emitter co-positioned with the center of the receivers matrix, as shown in Figure 2.5 for the case of the MESA Imaging SR4000. Indeed the sum of all the IR signals emitted by this configuration can be considered as a spherical wave emitted by a single emitter, called *simulated emitter* (Figure 2.5), placed at the center of the emitters constellation.

The fact that the actual emitters arrangement of Figure 2.5 is only an approximation of the non-feasible juxtaposition of single ToF sensor devices with emitter and receiver perfectly co-positioned introduces artifacts, among which a systematic distance measurement offset larger for the closer than for the further scene points. Figure 2.6 shows the actual emitters distribution of the MESA Imaging SR4000.

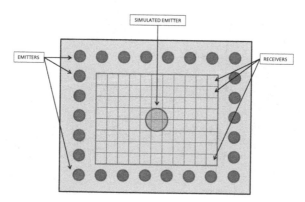

Fig. 2.5 Scheme of a matricial ToF camera sensor. The CCD/CMOS matrix of lock-in pixels is in red. The emitters (blue) are distributed around the lock-in pixels matrix and mimic a simulated emitter co-positioned with the center of the lock-in pixel matrix (light blue).

Fig. 2.6 The emitters of the MESA Imaging SR4000 are the red LEDs.

2.3.2 ToF Camera imaging characteristics

ToF cameras can be modeled as pin-hole imaging systems since their structure, schematically shown in Figure 2.7, similarly to standard cameras, has two major components, namely the sensor made by a $N_R \times N_C$ matrix of lock-in pixels as explained in Section 2.3.1 and the optics.

ToF cameras, differently from standard cameras, have also a third important component, namely an IR emitters set, typically placed near the optics as shown in Figure 2.7. Figure 2.7 also shows that the IR signal sent by the emitters set travels toward the scene (blue arrow), it is reflected by the different scene portions, it travels back to the camera and through the optics (red arrow) it is finally received by the different lock-in pixels of the ToF sensor. The signaling process shown by Figure 2.7 is the basis of the relationship between the various scene portions and the respective

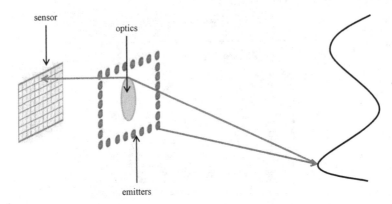

Fig. 2.7 ToF camera structure and signaling: propagation towards the scene (blue arrow), reflection (from the black surface on the right), back-propagation (red arrow) towards the camera through the optics (green) and reception (red sensor).

sensor pixels.

All the pin-hole imaging system notation and concepts introduced in Section 1.2 apply to ToF cameras. The notation will be used with pedix T in order to recall that it refers to a ToF camera. The CCS of the ToF camera will be called the

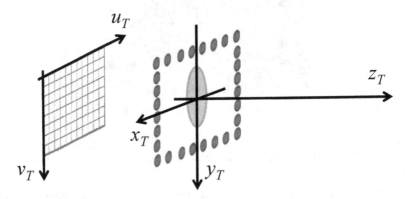

Fig. 2.8 2D T-reference system (with axes $u_T - v_T$) and 3D T-reference system (with axes $x_T - y_T - z_T$).

$3D - T$ *reference system.* The position of a scene point with respect to the $3D - T$ reference system will be denoted as P_T and its coordinates as $\mathbf{P}_T = [x_T, y_T, z_T]^T$. Coordinate z_T of P_T is called the *depth* of point P_T and the z_T-axis is called *depth axis.* The coordinates of a generic sensor pixel p_T of lattice Λ_T with the respect to the 2D-T reference system are represented by vector $\mathbf{p}_T = [u_T, v_T]^T$, with $u_T \in [0, ..., N_C]$ and $v_T \in [0, ..., N_R]$. Therefore the relationship between the 3D coordinates $\mathbf{P}_T = [x_T, y_T, z_T]^T$ of a scene point P_T and the 2D coordinates $\mathbf{p}_T = [u_T, v_T]^T$

of the pixel p_T receiving the IR radiation reflected by P_T is given by the perspective projection equation rewritten next for clarity's sake

$$z_T \begin{bmatrix} u_T \\ v_T \\ 1 \end{bmatrix} = K_T \begin{bmatrix} x_T \\ y_T \\ z_T \end{bmatrix} \qquad (2.15)$$

where the ToF camera intrinsic parameters matrix K_T is defined as in (1.7).

Because of lens distortion, coordinates $\mathbf{p_T} = [u_T, v_T]^T$ of (2.15) are related to the co-ordinates $\mathbf{\hat{p}_T} = [\hat{u}_T, \hat{v}_T]^T$ actually measured by the system by a relationship of type $\mathbf{\hat{p}_T} = [\hat{u}_T, \hat{v}_T]^T = \Psi(\mathbf{p_T})$, where $\Psi(\cdot)$ is a distortion transformation as described in Section 1.2.2. Model (1.12) supported by the camera calibration procedure [15] is widely used also with ToF cameras, as well as other more complex models, e.g., [16].

As already explained, each sensor pixel p_T directly estimates the radial distance $\hat{r}_T$ from its corresponding scene point P_T. With minor and neglectable approximation due to the non-perfect localization between emitters, pixel p_T and $3D - T$ reference system origin, the measured radial distance $\hat{r}_T$ can be expressed as

$$\hat{r}_T = \sqrt{\hat{x}_T^2 + \hat{y}_T^2 + \hat{z}_T^2} = \left\| \left[\hat{x}_T^2, \hat{y}_T^2, \hat{z}_T^2 \right]^T \right\|_2 \qquad (2.16)$$

From radial distance $\hat{r}_T$ measured at pixel p_T with distorted coordinates $\mathbf{\hat{p}_T} = [\hat{u}_T, \hat{v}_T]^T$ the $3D$ coordinates of $\mathbf{P}_T$ can be computed according to the following steps:

1. Given the lens distortion parameters, estimate the non-distorted $2D$ coordinates $\mathbf{p}_T = [u_T, v_T]^T = \Psi^{-1}(\mathbf{\hat{p}_T})$, where $\Psi^{-1}(\cdot)$ is the inverse of $\Psi(\cdot)$;
2. The value $\hat{z}_T$ can be computed from 2.15 and 2.16 as

$$\hat{z}_T = \frac{\hat{r}_T}{\left\| K_T^{-1} [u_T, v_T, 1]^T \right\|_2} \qquad (2.17)$$

   where K_T^{-1} is the inverse of K_T;
3. The values $\hat{x}_T$ and $\hat{y}_T$ can be computed by inverting (2.15), i.e., as

$$\begin{bmatrix} \hat{x}_T \\ \hat{y}_T \\ \hat{z}_T \end{bmatrix} = K_T^{-1} \begin{bmatrix} u_T \\ v_T \\ 1 \end{bmatrix} \hat{z}_T \qquad (2.18)$$

The operation of a ToF camera as imaging system can be summarized as follows. Each ToF camera sensor pixel, at each period of the modulation sinusoid, collects four samples s_R^0, s_R^1, s_R^2 and s_R^3 of the IR signal reflected by the scene. Every N periods of the modulation sinusoid, where N is a function of the integration time, each ToF sensor pixel estimates an amplitude value $\hat{A}$, an intensity value $\hat{B}$, a phase value $\widehat{\Delta\phi}$, a radial distance value $\hat{r}_T$ and the $3D$ coordinates $\mathbf{\hat{P}}_T = [\hat{x}_T, \hat{y}_T, \hat{z}_T]^T$ of the corresponding scene point.

Since amplitude $\hat{A}$, intensity $\hat{B}$ and depth $\hat{z}_T$ are estimated at each sensor pixel, ToF cameras handle them in matricial structures, and return them as 2D maps. Therefore a ToF camera, every N periods of the modulation sinusoid (which certainly correspond to several tens of times per second), provides the following types of data:

- An *amplitude map* $\hat{A}_T$, i.e., a matrix obtained by juxtaposing the amplitudes estimated at all the ToF sensor pixels. It is defined on lattice Λ_T and its values, expressed in volt $[V]$, belong to the pixel non-saturation interval. Map $\hat{A}_T$ can be modeled as realization of a random field $\mathscr{A}_T$ defined on Λ_T, with values (expressed in volt $[V]$) in the pixel non-saturation interval.

- An *intensity map* $\hat{B}_T$, i.e., a matrix obtained by juxtaposing the intensity values estimated at all the ToF sensor pixels. It is defined on lattice Λ_T and its values, expressed in volt $[V]$, belong to the pixel non-saturation interval. Map $\hat{B}_T$ can be modeled as realization of a random field $\mathscr{B}_T$ defined on Λ_T, with values (expressed in volt $[V]$) in the pixel non-saturation interval.

- A *depth map* $\hat{Z}_T$, i.e, a matrix obtained by juxtaposing the depth values estimated at all the ToF sensor pixels. It is defined on lattice Λ_T and its values, expressed in $[mm]$, belong to interval $\left[0, r_{MAX} = \frac{c}{2f_{mod}}\right)$. Map $\hat{Z}_T$ can be considered as realization of a random field $\mathscr{Z}_T$ defined on Λ_T, with values (expressed in $[mm]$) in $[0, r_{MAX})$.

By mapping amplitude, intensity and depth values to interval $[0, 1]$ the three maps $\hat{A}_T$, $\hat{B}_T$ and $\hat{Z}_T$ can be represented as images as shown in Figure 2.9 for a sample scene. For the scene of Figure 2.9 images $\hat{A}_T$ and $\hat{B}_T$ are very similar because the scene illumination is rather constant.

Fig. 2.9 Example of $\hat{A}_T$, $\hat{B}_T$ and $\hat{Z}_T$ (in this order from left to right in the figure).

2.3.3 Practical imaging issues

As expected the actual imaging behavior of ToF cameras is more complex than that of a simple pin-hole system and some practical issues must be taken into account.

First of all, it is not true that a sensor pixel is associated to a single scene point, but it is associated to a finite scene area, as shown in Figure 2.10. For this reason, each pixel receives the radiation reflected from all the points of the corresponding

Fig. 2.10 Finite size scene area (blue) associated to a ToF sensor pixel (red).

scene area. If the scene area is a flat region with somehow constant reflectivity, the approximation that there is a single scene point associated to the specific pixel does not introduce any artifact. However, if the area crosses a reflectivity discontinuity, the values of $\hat{A}_T(p_T)$ and $\hat{B}_T(p_T)$ estimated by the correspondent pixel p_T average somehow its different reflectivity values. A worse effect occurs if the area associated to p_T crosses a depth discontinuity. In this case assume that a portion of the area is at closer depth, called z_{near}, and another portion at further depth, called z_{far}. The resulting depth estimate $\hat{Z}_T(p_T)$ is a convex combination of z_{near} and z_{far}, where the combination coefficients depend on the percentage of area at z_{near} and at z_{far} respectively reflected on p_T. The pixels associated to such depth estimates are commonly called *flying pixels*. The presence of flying pixels leads to severe depth estimation artifacts, as shown by the example of Figure 2.11.

Multi-path propagation is a major interference in ToF camera imaging. As shown

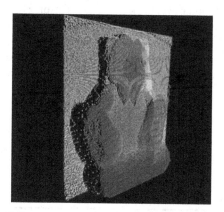

Fig. 2.11 An example of flying pixels at the depth edge between object and wall.

in Figure 2.12, an optical ray (red) incident to a non-specular surface is reflected in multiple directions (green and blue), a phenomenon commonly called *scattering*. The ideal propagation scenario of Figure 1.2, with co-positioned emitters and receivers, considers only the presence of the green ray of Figure 2.12, i.e., the ray

back reflected in the direction of the incident ray and disregards the presence of
the other (blue) rays. In practical situations, however, the presence of the other rays
may not always be neglectable. In particular, the ray specular to the incident ray
direction with respect to the surface normal at the incident point (thick blue ray) is
generally the reflected ray with greatest radiometric power. All the reflected (blue)

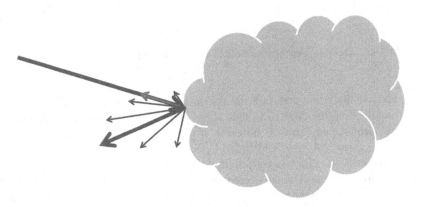

Fig. 2.12 Scattering effect.

rays may first hit others scene points and then travel back to the ToF sensor, affecting
therefore the distance measurements of other scene points. For instance, as shown
in Figure 2.13, an emitted ray (red) may be firstly reflected by a point surface (A)
with a scattering effect. One of the scattered rays (orange) may then be reflected by
another scene point (B) and travel back to the ToF sensor. The distance measured by
the sensor pixel relative to B is therefore a combination of two paths, namely path
to ToF camera - B - ToF camera and path ToF camera-A-B-ToF camera. The coefficients of such a combination depend on the optical amplitude of the respective rays.
Since the radial distance of a scene point P from the ToF camera is computed from

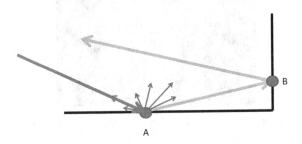

Fig. 2.13 Multi-path phenomenon: the incident ray (red) is reflected in multiple directions (blue
and orange rays) by the surface at point A. The orange ray reaches then B and travels back to the
ToF sensor.

the time-length of the shortest path between P and the ToF camera, the multi-path effect leads to over-estimate the scene points distances.

Multi-path is one of the major error sources of ToF cameras distance measurements. Since multi-path is scene dependent it is very hard to model. Currently there is no method for its compensation, but there are practical provisions that might alleviate the multi-path effects, as explained in [17].

2.4 Conclusion and further reading

This chapter introduces the basic operation principles of continuous wave ToF sensors. A comprehensive review of current ToF technology, not only of CW type, is given by in [18]. A classical and accurate description of CW ToF operation and technology is given by [9]. In depth analysis of ToF noise sensor sources and practical provisions against noise can be found in [12, 13, 10, 11]. ToF cameras image formation can be approximated by the pinhole model, typical of standard cameras, recalled in Section 1.2. More extensive treatments on topics such as image formation, projective geometry and camera calibration can be found in [19, 20, 15, 16]. More on ToF camera calibration will be seen in Chapter 4.

ToF cameras can be considered as special Multiple Input - Multiple Output (MIMO) communication systems, where the emitters array is the input array and the lock-in matrix of the ToF sensor the output array. This kind of framework in principle would allow to approach multi-path as customarily done in communication systems. However the number of input and output channels of a ToF camera (for the MESA SR4000 there would be 24 input channels, associated to the emitters, and 176×144 output channels associated to the lock-in pixels of the sensor) is way superior to the complexity of the MIMO systems used in telecommunications (where the number of inputs and outputs rarely exceeds the 10 units). The current multi-path analysis methods used for MIMO systems cannot be applied to ToF cameras, however the application of communications systems techniques for characterizing ToF cameras operations and improving their performances is an attractive possibility.

References

1. "Mesa imaging." http://www.mesa-imaging.ch.
2. "Pmd technologies." http://www.pmdtec.com/.
3. "Softkinetic." http://www.softkinetic.com/.
4. "Microsoft®." http://www.microsoft.com.
5. "Panasonic d-imager." http://www.panasonic-electric-works.com.
6. "Iee." http://www.iee.lu.
7. "Csem." http://www.csem.ch.
8. "Fbk." http://www.fbk.eu.
9. R. Lange, *3D Time-Of-Flight distance measurement with custom solid-state image sensors in CMOS/CCD-technology*. PhD thesis, University of Siegen, 2000.

10. B. Buttgen, T. Oggier, M. Lehmann, R. Kaufmann, and F. Lustenberger, "Ccd/cmos lock-in pixel for range imaging: Challenges, limitations and state-of-the-art," in *1st range imaging research day*, 2005.

11. F. Muft and R. Mahony, "Statistical analysis of measurement processes for time-of flight cameras," *Proceedings of SPIE the International Society for Optical Engineering*, 2009.

12. B. Buttgen and P. Seitz, "Robust optical time-of-flight range imaging based on smart pixel structures," *Circuits and Systems I: Regular Papers, IEEE Transactions on*, vol. 55, pp. 1512 –1525, july 2008.

13. T. Kahlmann and H. Ingensand, "Calibration and development for increased accuracy of 3d range imaging cameras," *Journal of Applied Geodesy*, vol. 2, pp. 1–11, 2008.

14. C. Uriarte, B. Scholz-Reiter, S. Ramanandan, and D. Kraus, "Modeling distance nonlinearity in tof cameras and correction based on integration time offsets," in *Progress in Pattern Recognition, Image Analysis, Computer Vision, and Applications*, Springer Berlin / Heidelberg, 2011.

15. J. Heikkila and O. Silven, "A four-step camera calibration procedure with implicit image correction," in *CVPR*, 1997.

16. D. Claus and A. Fitzgibbon, "A rational function lens distortion model for general cameras," in *CVPR*, 2005.

17. M. Imaging, "Sr4000 user manual." http://www.mesa-imaging.ch.

18. D. Stoppa and F. Remondino, eds., *TOF Range-Imaging Cameras*. Springer, 2012.

19. R. Szeliski, *Computer Vision: Algorithms and Applications*. New York: Springer, 2010.

20. R. Hartley and A. Zisserman, *Multiple View Geometry in Computer Vision*. Cambridge University Press, 2004.

Chapter 3
Microsoft Kinect^{TM} Range Camera

The Microsoft Kinect^{TM} range camera (for simplicity just called Kinect^{TM} in the sequel) is a light-coded range camera capable to estimate the 3D geometry of the acquired scene at 30 fps with VGA (640×480) spatial resolution. Besides its light-coded range camera, the Kinect^{TM} also has a color video-camera and an array of microphones. In the context of this book the Kinect^{TM} light-coded range camera is the most interesting component, and the name Kinect^{TM} will be often referred to it rather than to the whole product.

From the functional point of view the Kinect^{TM} range camera is very similar to the ToF cameras introduced in the previous chapter, since they both estimate the 3D geometry of dynamic scenes, but current Kinect^{TM} technology is totally different. The Kinect^{TM} range camera is based on the Primesensor^{TM} chip produced by Primesense [1]. Although the Kinect^{TM} (Figure 3.1) is the most popular consumer electronic product based on such a chip, it is not the only one. Asus X-tion Pro and X-tion Pro Live [2] are other products with a range camera based on the same chip. All these range cameras, as it will be seen, support an IR video-camera and an IR

Fig. 3.1 Microsoft Kinect^{TM}.

projector projecting on the scene IR light coded patterns in order to obtain a matricial implementation of the active triangulation principle for estimating scene depth and 3D geometry.

This chapter firstly describes how active triangulation can be implemented in a matricial form by means of different light coding techniques. The general operation of Kinect^{TM} can be inferred from that of light-coding systems, of which the Kinect^{TM} is a special case, but its implementation details are still undisclosed. Nevertheless some characteristics of its functioning can be derived from its operation as reported by some reverse engineering studies [3, 4].

The practical imaging issues responsible for the Kinect^{TM} error sources and data characteristics are finally considered in the last section of this chapter, as previously done for ToF cameras.

3.1 Matricial active triangulation

The Kinect^{TM} range camera has the structure shown in Figure 1.7, reported also in Fig. 3.2, with a camera C and a projector A, and in principle it implements active triangulation. Namely, given a point p_C in the image acquired by C and its conjugate point p_A in the pattern projected by A, the depth z of the 3D point P associated to p_C with respect to the C-3D reference system is computed by active triangulation as described in Section 1.2.4.

In order to use a pattern invisible to human eyes, both projector and camera operate at IR wavelengths. The data made available from the Kinect^{TM} range camera at $30\,[fps]$ are:

- the image acquired by C, that in the Kinect^{TM} case is an IR image called I_K and it is defined on the lattice Λ_K associated to the C sensor. The axes that identify Λ_K coincide with u_C and v_C of Figure 1.7. The values of I_K belong to interval $[0,1]$. Image I_K can be considered a realization of a random field $\mathcal{I}_K$ defined on Λ_K, with values in $[0,1]$.
- The estimated disparity map, called $\hat{D}_K$ is defined on the lattice Λ_K associated to the C sensor. The values of $\hat{D}_K$ belong to interval $[d_{min}, d_{max}]$, where d_{min} and d_{max} are the minimum and maximum allowed disparity values. Disparity map $\hat{D}_K$ can be considered a realization of a random field $\mathcal{D}_K$ defined on Λ_K, with values in $[d_{min}, d_{max}]$.
- The estimated depth map computed by applying (1.14) to $\hat{D}_K$, called $\hat{Z}_K$ is defined on the lattice Λ_K associated to the C sensor. The values of $\hat{Z}_K$ belong to the interval $[z_{min}, z_{max}]$, where $z_{min} = \frac{bf}{d_{max}}$ and $z_{max} = \frac{bf}{d_{min}}$ are the minimum and maximum allowed depth values respectively. Depth map $\hat{Z}_K$ can be considered as a realization of a random field $\mathcal{Z}_K$ defined on Λ_K, with values in $[z_{min}, z_{max}]$.

The spatial resolution of I_K, $\hat{D}_K$ and $\hat{Z}_K$ is 640×480. The minimum measurable depth is $0.5\,[m]$ and the nominal maximum depth is$15\,[m]$. According to [3], the values of b and f are $75\,[mm]$ and $585.6\,[pxl]$ respectively. Therefore the minimum allowed disparity is $2\,[pxl]$ and the maximum is $88\,[pxl]$. Figure 3.3 shows an example of I_K, $\hat{D}_K$ and $\hat{Z}_K$ acquired by the Kinect^{TM} range camera.

Section 1.2 introduced active triangulation applied to single pixels p_C of the $N_R \times N_C$ images I_K acquired by camera C. The Kinect^{TM}, instead, simultaneously

Scene

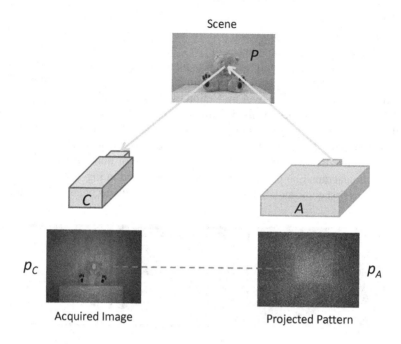

p_C

Acquired Image

p_A

Projected Pattern

Fig. 3.2 Matricial active triangulation flow: pixel p_A (green dot) is coded in the pattern. The pattern is projected to the scene and acquired by C. The 3D point associated to p_A is P and the conjugate point of p_A (green dot) in I_K is p_C (blue dot). The correspondence estimation algorithm (red dashed arrow) estimates the conjugate points.

applies active triangulation to all the $N_R \times N_C$ pixels of I_K, a procedure called *matricial active triangulation*, which requires a number of special provisions discussed in the next subsection.

The major difficulty with matricial active triangulation is keeping the correspondence problem as simple as in the single point case seen in Section 1.2. This issue can be handled by designing the patterns projected by A by the *light coding* methods described next. The specific design of the light-coded pattern is the actual core of the Kinect™ range camera. The next two subsections present current light coding techniques and give some hints about the specific Kinect™ operation.

3.1.1 Light Coding Techniques

Let us assume that the projected pattern has $N_R^A \times N_C^A$ pixels p_A^i, $i = 1,...,N_R^A \times N_C^A$ where N_R^A and N_C^A are the number of rows and columns of the projected pattern respectively. In order to apply active triangulation, each pixel needs to be associated to a *code-word*, i.e., a specific local configuration of the projected pattern. The pattern undergoes projection by A, reflection by the scene and capture by C (see Figure 3.2). A correspondence estimation algorithm analyzes the received code-words in the acquired images I_K in order to compute the conjugate of each pixel of the projected pattern. The goal of pattern design (i.e., code-words selection) is to adopt code-words effectively decodable even in presence of non-idealities of the pattern projection/acquisition process, as pictorially indicated in Fig. 3.2 .

Let us first consider what makes a code-word highly decodable. It is intuitive that

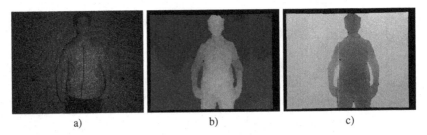

a)   b)   c)

Fig. 3.3 Example of I_K, $\hat{D}_K$ and $\hat{Z}_K$ acquired by the Kinect^{TM} range camera.

the more the code-words are different the more robust is the coding against disturbances and self-interferences. For a given cardinality of the total number of possible code-words, the smaller is the number of used code-words, the greater become the differences between the various code-words and the more robust is the code. Since for a calibrated and rectified setup conjugate points lie on horizontal lines, the coding problem can be independently formulated for each row in order to keep as low as possible the cardinality of the total number of possible code-words. Assume that for each row of the projected pattern there are $N = N_C^A$ pixels $p_A^1, p_A^2,...,p_A^N$ to be encoded with N code-words $w_1, w_2,...,w_N$. Each code-word is represented by a specific local pattern distribution. Clearly, the more the local pattern distribution of a single pixel differs from the local pattern distribution of the other pixels of the same row, the more robust will be the coding.

A code-words alphabet can be implemented by a light projector considering that it can produce n_P different illumination values called *pattern primitives* (e.g., $n_P = 2$ for a binary black-and-white projector, $n_P = 2^8$ for a $8 - bit$ gray-scale projector and $n_P = 2^{24}$ for a RGB projector with $8 - bit$ color channels). The local distribution of a pattern for a pixel p_A is given by the illumination values of the pixels in a window around p_A. If the window has n_W pixels, there are $n_P^{n_W}$ possible pattern configurations on it. From the set of all possible configurations, N configurations need to be chosen as code-words. What is projected to the scene and acquired by C is the pattern resulting from the code-words relative to all the pixels of the projected pattern.

Let us recall that, because of the geometrical properties of a calibrated and rectified system described in Section 3.1, a pixel p_A of the pattern, with coordinates $\mathbf{p_A} = [u_A, v_A]^T$, is projected to the scene point P, with coordinates $\mathbf{P} = [x, y, z]^T$, and acquired by C in p_C, with coordinates $\mathbf{p_C} = [u_A + d, v_A]^T$. The projection/acquisition process introduces an horizontal shift d proportional to the inverse of the depth z of P according to (1.14). Disparity shift d is the most important quantity in the active triangulation process, and it needs to be accurately estimated since it carries the 3D geometry information relative to the considered scene point P (obtainable from (1.15)). The disparity values in matricial active triangulation are handled as arrays, like the disparity map $\hat{D}_K$ of Kinect^{TM} introduced in Section 3.1.

In the pattern projection/acquisition process, there are a number of factors transforming the projected pattern and introducing artifacts which must be taken into consideration:

a) *Perspective distortion.* Since the scene points may have different depth value z, neighboring pixels of the projected pattern may not be mapped to neighboring pixels of I_C. In this case the local distribution of the acquired pattern becomes a distorted version of the relative local distribution of the projected pattern.

b) *Color or gray-level distortion due to scene color distribution and reflectivity properties of the acquired objects.* The projected pattern undergoes reflection (and absorption) by the scene surfaces. The ratio between incident and reflected radiant power is given by the scene reflectance, which is generally related to the scene color distribution. In particular in the case of IR light, used by the Kinect^{TM} projector, the appearance of the pixel p_C on the camera depends on the reflectance of the scene surface at the IR frequency used by the projector. An high intensity pixel of the projected pattern at p_A for instance may undergo a strong absorption because of the low reflectance value of the scene point to which it is projected, and the values of its conjugate pixel p_C on I_K may consequently appear much darker. This is a very important issue, since it might completely distort the projected code-words. The second row of Figure 3.4 shows how the radiometric power of the projected pattern is reflected by surfaces of different color.

c) *External illumination.* The color acquired by the color camera depends on the light falling on the scene surfaces, which is the sum of the projected pattern and of the scene illumination (i.e. sunlight, artificial light sources, etc..). This second contribution with respect to code-word detection acts as a noise source added to the information signal of the projected light.

d) *Occlusions.* Because of occlusions, not all the pattern pixels are projected to 3D points seen by C. Depending on the 3D scene geometry, in general between the pattern pixels and the pixels of I_K (the image acquired by C), there may not be a biunivocal association. It is important therefore to correctly identify the pixels of I_K that do not have a conjugate point in the pattern, in order to discard erroneous correspondences.

e) *Projector and camera non-idealities.* Both projector and camera are not ideal imaging systems. In particular, they generally do not behave linearly with respect to the projected and the acquired colors or gray-levels.

f) *Projector and camera noise.* The presence of random noise in the projection and acquisition processes is typically modeled as Gaussian additive noise in the acquired images I_K.

Figure 3.4 shows some examples of the considered transformations/distortions. As result of the above transformations/distortions, an acquired code-word may be very different from the projected one and due to occlusions some pixels of I_K may even not correspond to any pixel of the projected pattern.

In order to understand how to possibly mitigate such potentially disruptive effects for the correspondence problem by comparison methods assisted by suitable patterns, let us make two considerations. The first one is that the correspondences estimation process is characterized by two fundamental decisions, namely:

• what code-word assign to each pixel p_A of the projected pattern. The code-word corresponds to a pattern to be projected on a window centered at p_A (all the local pattern distributions of neighboring pixels are fused together in a single projected pattern);
• what code-word assign to each pixel p_C of I_K, or equivalently how to detect the code-word most similar to the local pattern distribution around p_C in order to correctly identify the conjugate pixel p_C of the projected pattern pixel p_A.

The second consideration is that there are mainly three possible coding schemes, pictorially exemplified in Figure 3.5:

a) *Direct coding*: the code-word associated to each pixel p_A is represented by the pattern value at the pixel itself (i.e., the gray-level or the color of the pattern at p_A). In this case there may be up to n_P code-words, since $n_W = 1$. Therefore the maximum number of pattern columns to decode is $N_C = n_P$.

b) *Time-multiplexing coding*: a sequence of T patterns is projected to the surface to be measured at T subsequent times. The code-word associated to each pixel p_A is the sequence of the T pattern values (i.e., of gray-level or color values) at pixel p_A. In this case there may be up to n_P^T code-words and the maximum number of pattern columns to decode is $N_C = n_P^T$.

c) *Spatial-multiplexing coding*: the code-word associated to each pixel p_A is the spatial pattern distribution in a window of n_W pixels centered around p_A. In this case there might be up to $n_P^{n_W}$ code-words. For instance if the window has 9 rows and 9 columns, which is a common choice, $n_W = 81$. It is important to note how in this case neighboring pixels share parts of their code-words, thus making their coding interdependent.

Each one of the considered coding strategies have different advantages and disadvantages as described in [5], which gives an exhaustive review of all these techniques. In particular, direct coding methods are the easiest to implement and allow

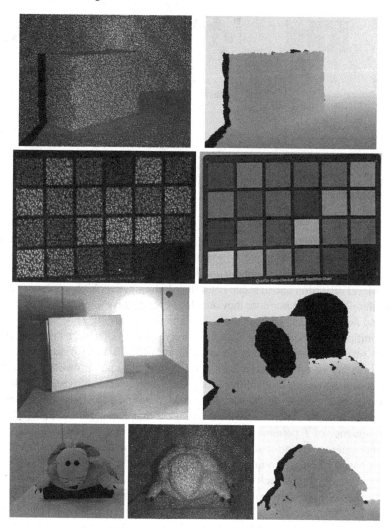

Fig. 3.4 Examples of different artifacts affecting the projected pattern (*in the depth maps black pixels correspond to locations without a valid depth measurement*). *First row*: projection of the IR pattern on a slanted surface and corresponding depth map; observe how the pattern is shifted when the depth values change and how perspective distortion affects the pattern on the slanted surfaces. *Second row*: Kinect™ pattern projected on a color checker and corresponding depth-map; observe how the pattern appearance depends also on the surface color. *Third row*: a strong external illumination affects the acquired scene; the acquired IR image saturates in correspondence of the strongest reflections and the Kinect™ is not able to acquire the depth of those regions. *Fourth row*: the occluded area behind the ear of the teddy bear is visible from the camera but not from the projector viewpoint; consequently the depth of this region can not be computed.

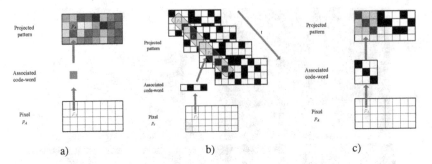

Fig. 3.5 Examples of coding strategies: a) direct coding; b) time-multiplexing coding; c) spatial-multiplexing coding.

for the capture of dynamic scenes, since they require the projection of a single pattern. They have the advantage to be potentially able to deal with occlusions as well as with projective distortion, and the disadvantage to be extremely sensitive to color or gray-level distortion due to scene color distribution, reflectivity properties and external illumination. Furthermore they are also very sensitive to projector and camera noise and non-idealities. Examples of these approaches are [6, 7].

Time-multiplexing coding allows to adopt a very small set of pattern primitives (e.g., a binary set), in order to create arbitrarily different code-words for each pixel. Such code-words are therefore robust with respect to occlusions, projective distortion, projector and cameras non-idealities and noise. They are also able to deal with color and gray-level distortion due to scene color distribution, reflectivity properties and external illumination. Their major disadvantage is that they require the projection of a time sequence of T patterns for a single depth measurement, hence their application is not suited to dynamic scenes. Examples of time-multiplexing approaches are the ones based on binary coding [8] and gray coding with phase-shifting [9].

The spatial-multiplexing techniques are the most interesting for the acquisition of dynamic scenes since they require the projection of a single pattern, like direct coding methods. They are generally robust with respect to projector and cameras non-idealities. They can also handle noise and color or gray-level distortion due to scene color distribution, reflectivity properties and external illumination. Difficulties come with occlusions and perspective distortion, because of the finite size of the pattern window associated to each code-word. The choice of the window size n_W associated to each code-word is crucial. Indeed, on one hand, the smaller is n_W the more robust is the coding with respect to perspective distortion, since it is more likely that all the scene points on which the pattern window is projected share the same disparity. On the other hand, the greater is n_W the greater is the robustness with respect to projector and cameras non-idealities and noise or color distortion. Examples of spatial-multiplexing coding are the ones based on non-formal codification [10], the ones based on De Bruijns sequences [11] and the ones based on M-arrays [12].

3.1.2 About Kinect^{TM} range camera operation

3.1.2.1 Horizontal uncorrelation of the projected pattern

The Kinect^{TM} is a proprietary product and its algorithms are still undisclosed, however some characteristics of the adopted light coding system can be deduced from
its operation [13, 3, 4].

The Kinect^{TM} range camera adopts a spatial-multiplexing approach that allows
to robustly capture dynamic scenes at high frame-rate ($30\,[fps]$). The spatial resolution of the output depth maps is 640×480 pixels, although the camera C and projector A probably have a different internal resolution still undisclosed. The projected
pattern, a picture of which is shown in Figure 3.6, is characterized by an uncorre-
lated distribution across each row [13]. This means that the covariance between the

Fig. 3.6 Picture of the pattern projected to a planar surface acquired by a high-quality photo-
camera.

spatial-multiplexing window centered at a given pattern point p_A^i, with coordinates
$\mathbf{p}_A^i = [u_A^i, v_A^i]^T$ and the projected pattern on the spatial multiplexing windows centered at point p_A^j with coordinates $\mathbf{p}_A^j = [u_A^j, v_A^j]^T$, assuming $v_A^i = v_A^j$, is 0 if $i \neq j$ and
1 if $i = j$. More precisely, if $s(u,v)$ denotes the projected pattern and $W(u_A, v_A)$ the
support of the spatial multiplexing window centered at p_A, the above property can
be written as

$$C(u^i,u^j,v^i) = \sum_{(u,v)\in W(u_A,v_A)}[s(u-u_A^i,v-v_A^i)-\overline{s}(u_A^i,v_A^i)]\cdot[s(u-u_A^j,v-v_A^i)-\overline{s}(u_A^j,v_A^i)]$$

$$= \begin{cases} 1 \ i=j \\ 0 \ i\neq j \end{cases} \tag{3.1}$$

where $\overline{s}(u_A,v_A) = \sum_{(u,v)\in W} s(u-u_A,v-v_A)$ is the average of the spatial multiplexing window on W centered at $(u_A,v_A)^T$ and $C(u^i,u^j,v^i)$ is the horizontal covariance of the projected pattern between supports centered at p_A^i and p_A^j.

Some reverse engineering analysis suggest 7×7 [3] or 9×9 [4] as support of the spatial multiplexing window adopted by Kinect^{TM}.

It is important to note that (3.1) strictly holds only for the covariance of the projected ideal pattern. The patterns actually measured by the Kinect^{TM} range camera, as expected, are affected by all the already described transformations/distortions among which the non-idealities of camera and projector. The pattern acquired by the Kinect^{TM} IR camera (in front of a flat surface) is shown in Figure 3.7. Figure 3.8 shows its covariance with $p_A^i = [19,5]^T$ and $p_A^j = [u,5]^T$, $u \in \{1,2,...,200\}$ and a 9×9 support W. Figure 3.8 shows a plot of $C(19,u,5)$ as defined in (3.1) versus u and, in spite the acquired pattern is a distorted version of the projected one, it exhibits a very clear peak at $u = 19$ where it is expected.

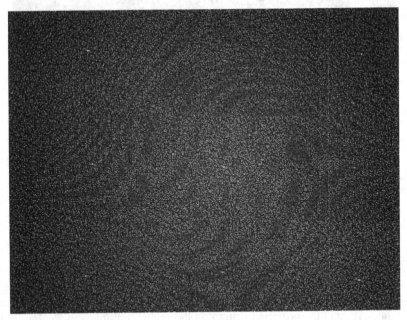

Fig. 3.7 Pattern acquired by the Kinect^{TM} IR camera.

The horizontal uncorrelation of the projected pattern is common to a variety of spatial multiplexing methods, such as [11, 14], since it is rather useful for the effective solution of the correspondences problem.

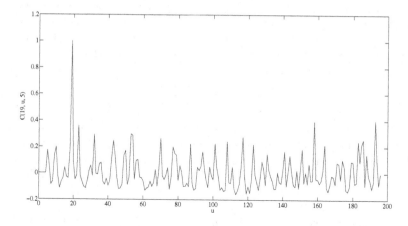

Fig. 3.8 Covariance of the pattern acquired by the Kinect^{TM} for p_A^i with coordinates $\mathbf{p}_A^i = [19,5]^T$ and $p_A^j = [u,5]^T$ and $u \in \{1,...,200\}$.

Indeed, in this case, for a given pixel p_C^j of I_K and for all the pixels p_A^i of the same row in the projected pattern, the covariance ideally presents a unique peak in correspondence of the actual couple of conjugate points, provided the introduced distortions and non-idealities are moderately disruptive. In a number of real situations, conjugate points can be simply detected from the covariance maximum between a window centered around the specific pixel p_C in I_K and all the possible conjugates p_A^i in the same row of the projected pattern, as shown by the example of Figure 3.8. The covariance maximum is not well-defined when there are no conjugate points because of occlusions or when the impact of one of the above mentioned sources of distortion is too strong and the acquired code-words in I_K are too distorted to give significant auto-covariance values.

3.1.2.2 Rerence image and correspondence detection

The direct comparison between the images I_K acquired by C with the patterns projected by A would be affected by all the non-idealities of both systems. Such limitation can be overcome by means of a calibration procedure schematically shown in Figure 3.9.

The procedure requires the off-line acquisition in absence of background illumination of a highly reflective surface oriented orthogonally to the optical axis of the camera C, positioned at a known distance from C. In this case the acquired surface is characterized by a constant and known distance z_{REF} and therefore by a constant and known disparity d_{REF} that can be computed by (1.14). The acquired image is called *reference image* [13].

In any subsequent generic scene acquisition, it is possible to associate each point

p_{REF} with coordinates $\mathbf{p}_{REF} = [u_{REF}, v_{REF}]^T$ of the reference image to a point p of the acquired image and express its coordinates with respect to those of p_{REF}, as $\mathbf{p} = [u, v]^T = [u_{REF} + d_{REL}, v_{REF}]$. In this way the actual disparity value d_A of each scene point can be computed by adding d_{REF} to the relative disparity value d_{REL} directly computed from the acquired image

$$d_A = d_{REF} + d_{REL} \tag{3.2}$$

The usage of such a reference image allows to overcome a number of difficulties induced by the various transformations distorting the acquired pattern with respect to the projected one, among which, notably, the camera and projector non-idealities. In other words, the comparison of image I_K relative to a generic scene with the ref-

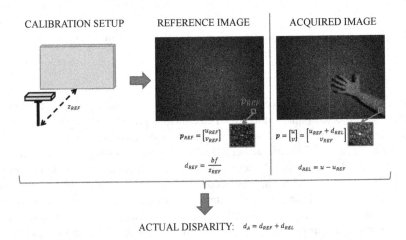

Fig. 3.9 Pictorial illustration of the reference image usage: (left) The Kinect™ in front of a flat surface at known distance z_{REF}; (middle) reference image and computation of d_{REF} from z_{REF}; (right) generic I_K with pixel coordinates referred to the reference image coordinates.

erence image (which is an acquired version of the projected pattern under known scene conditions) is an implicit way to avoid the non-idealities and distortions due to C and A.

Disparity estimation with respect to the reference image is a procedure very similar to the computational stereopsis between two rectified images [15, 16]. In this context estimating the conjugate pixels from the covariance maximum, as described above, is a *local algorithm*, i.e., a method which considers a measure of the local similarity (covariance) between all the pairs of possible conjugate points and simply selects the pair that maximizes it. As observed at the end of Section 1.2.3, there is a great number of more complex and way more effective computational stereopsis techniques that do not consider each couple of points on its own but exploit global optimization schemes. It is rather likely that the Kinect™, besides the reference image scheme, adopts a global method for computing the disparities.

3.1.2.3 Subpixel refinement

Once the disparity values of all the pixels of I_K are available, some further refinement is appropriate. In particular, the disparity values computed by covariance maximization or other techniques are implicitly assumed to be integers. It is well known from the stereo literature that this limits somehow the depth resolution and that these estimates can be improved by the so-called *sub-pixel refinement* techniques [15, 16], which are generally adopted in spite they increase the computational complexity. According to the analysis of [4], the Kinect^{TM} uses a sub-pixel refinement process with an interpolation factor of 8.

3.2 Practical imaging issues

As already explained in the previous section, there are multiple error sources potentially leading to non correct correspondence estimates, thus resulting in depth estimation errors. Some of these error sources are due to the camera and projector, some to the adopted correspondence estimation algorithm and some to the geometry of the acquired scene. From an experimental analysis of the data produced by the Kinect^{TM} range camera, the main depth estimation errors are due to:

- Low reflectivity and background illumination. In case of low reflectivity (e.g., the foot of the teddy bear of Figure 3.10a) and excessive background illumination (Figure 3.10b), the camera is unable to acquire any information about the reflected pattern and the correspondence estimation algorithm does not produce any result. Hence in these situations there is typically no depth information available.
- Excessively slanted surfaces. In this case (e.g., the table plane of Figure 3.10.a), the perspective distortion is so strong that in the spatial multiplexing window support there are too many pixels characterized by different disparities and the correspondence estimation algorithm is unable to give any result. Hence also in these situations there is no depth information available.
- Depth discontinuities and occlusions (Figure 3.10c). Near depth discontinuities, the spatial multiplexing window support may include pixels associated to different disparities, with consequent possibility of errors. Moreover, close to depth discontinuities, there are occlusions due to scene points on the furthest surface not visible from either the camera or the projector. Hence depth estimation is not feasible for these points. The Kinect^{TM} range camera however adopts some heuristics and delivers depth estimates also for these points by interpolating the depth values of neighbor pixels. Such heuristic assignments sometimes lead to misalignments between real and estimated depth discontinuities (up to 10 pixels).

Some examples of the artifacts due to the above issues are shown in Figure 3.10. The spatial resolution of stereo vision systems typically derives from that of the

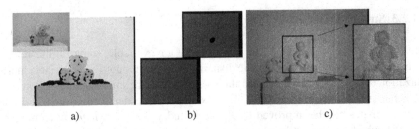

Fig. 3.10 Different depth estimation errors: a) low reflectivity (teddy bear) and tilted surfaces (table) issues (black pixels correspond to missing depth data); b) same flat surface acquired with (top) and without (bottom) background illumination; c) depth discontinuities and occlusion effects are visible on the baby's arm.

adopted sensor. In the case of the Kinect™ I_K, $\hat{D}_K$ and $\hat{Z}_K$ all have a spatial resolution coinciding with the sensor resolution of 640×480. As previously said, the actual internal resolutions of camera C and projector A are not known.

It is worth recalling another issue of depth estimates obtained by triangulation, i.e., it is possible to manipulate Equation (1.14) to show that the depth estimate resolution decreases with the square of the distance [17]. Therefore the quality of the depth measurements obtained by matricial active triangulation is worse for the furthest scene points than for the closest ones [3].

3.3 Conclusion and further reading

Originally introduced for gesture recognition in the gaming context [18], the Kinect™ has readily been adopted for several other purposes such in robotics [19] and 3D scene reconstruction [20]. Already just one year after its introduction, the number of applications based on Kinect™ is dramatically growing and its usefulness and relevance for 3D measurement purposes are becoming apparent.

The Kinect™ range camera is based on a Primesense proprietary light-coding technique and neither Microsoft nor Primesense have disclosed yet all the sensor implementation details. Several patents, among which [13], cover the technological basis of the range camera, and the interested reader might look at them or to current reverse engineering works [3, 4].

This chapter presents the Kinect™ range camera emphasizing its general operation principles rather than its implementation details, not only because the latters are not available, but also because the technology behind Kinect™ is likely to rapidly evolve. In a longer term perspective the general light coding operation principles are likely to be more useful than the characteristics of the first generation product. A comprehensive review of light-coding techniques can be found in [5]. The last section of this chapter presents the main characteristics of Kinect™ depth measurements.

Metrological studies about Kinect™ data are not available yet but they are expected soon in light of the great potential of Kinect™ as depth measurement instrument.

References

1. Primesense. http://www.primesense.com/.
2. Asus, "Xtion pro." http://www.asus.com/Multimedia/Motion_Sensor/Xtion_PRO/.
3. J. Smisek, M. Jancosek, and T. Pajdla, "3d with kinect," in *1st IEEE Workshop on Consumer Depth Cameras for Computer Vision*, 2011.
4. K. Konoldige and P. Mihelich, "Technical description of kinect calibration," tech. rep., Willow Garage, http://www.ros.org/wiki/kinect_calibration/technical, 2011.
5. J. Salvi, J. Pags, and J. Batlle, "Pattern codification strategies in structured light systems," *Pattern Recognition*, vol. 37, pp. 827–849, 2004.
6. B. Carrihill and R. Hummel, "Experiments with the intensity ratio depth sensor," *Computer Vision, Graphics, and Image Processing*, vol. 32, no. 3, pp. 337–358, 1985.
7. J. Tajima and M. Iwakawa, "3-d data acquisition by rainbow range finder," in *Proceedings of the 10th International Conference on Pattern Recognition*, 1990.
8. M. Trobina, "Error model of a coded-light range sensor," tech. rep., Communication Technology Laboratory Image Science Group, ETH-Zentrum, Zurich, 1995.
9. D. Bergmann, "New approach for automatic surface reconstruction with coded light," in *Proc. Remote Sensing and Reconstruction for Three-Dimensional Objects and Scenes, SPIE*, 1995.
10. M. Maruyama and S. Abe, "Range sensing by projecting multiple slits with random cuts," *IEEE Transactions on Pattern Analysis and Machine Intelligence*, vol. 15, pp. 647–651, 1993.
11. K. Boyer and A. Kak, "Color-encoded structured light for rapid active ranging," *IEEE Transactions on Pattern Analysis and Machine Intelligence*, vol. 9, pp. 14–28, 1987.
12. T. Etzion, "Constructions for perfect maps and pseudorandom arrays," *IEEE Transactions on Information Theory*, vol. 34, no. 5, pp. 1308–1316, 1988.
13. "Patent application us2010 0118123," 2010.
14. R. Morano, C. Ozturk, R. Conn, S. Dubin, S. Zietz, and J. Nissano, "Structured light using pseudorandom codes," *IEEE Transactions on Pattern Analysis and Machine Intelligence*, vol. 20, pp. 322–327, 1998.
15. R. Szeliski, *Computer Vision: Algorithms and Applications*. New York: Springer, 2010.
16. D. Scharstein and R. Szeliski, "A taxonomy and evaluation of dense two-frame stereo correspondence algorithms," *International Journal of Computer Vision*, 2001.
17. G. Bradski and A. Kaehler, *Learning OpenCV: Computer Vision with the OpenCV Library*. O'Reilly, 2008.
18. J. Shotton, A. Fitzgibbon, M. Cook, T. Sharp, M. Finocchio, R. Moore, A. Kipman, and A. Blake, "Real-Time Human Pose Recognition in Parts from Single Depth Images," in *CVPR*, 2011.
19. "Willow garage turtlebot." http://www.willowgarage.com/turtlebot.
20. S. Izadi, D. Kim, O. Hilliges, D. Molyneaux, R. Newcombe, P. Kohli, J. Shotton, S. Hodges, D. Freeman, A. Davison, and A. Fitzgibbon, "KinectFusion: real-time 3D reconstruction and interaction using a moving depth camera," in *Proceedings of the 24th annual ACM Symposium on User Interface Software and Technology*, 2011.

Chapter 4
Calibration

Color imaging instruments, such as photo and video-cameras, and depth imaging instruments, such as ToF cameras and the Kinect^{TM}, require a preliminary calibration in order to be used for measurement purposes. Calibration must usually account both for the internal characteristics and the spatial positions of the considered instruments, and it needs to be accurate and precise for meaningful measures.

The first part of this chapter formalizes calibration for generic measurement instruments in order to give a unified framework to the calibration of the imaging instruments considered in this book, i.e. standard cameras, ToF cameras, the Kinect^{TM} range camera and stereo vision systems. The Kinect^{TM} and the stereo vision systems, although made by a pair of devices, will be considered as single depth measurement instruments.

The second part of this chapter considers the calibration of heterogeneous imaging systems made by a standard camera and a ToF camera or a Kinect^{TM}, or by a stereo system and a ToF camera or a Kinect^{TM}, because of their great interest for the applications described in the subsequent chapters.

4.1 Calibration of a generic measurement instrument

Consider a generic instrument measuring one (or more) attribute a of an object. Let us denote with $\hat{a}$ the measured value and with a^* the actual quantity to be measured, taking values in interval $[a_{min}, a_{max}]$. Measurements are generally characterized by two types of errors:

- systematic errors, reducing measurements accuracy;
- stochastic errors, reducing measurements precision (repeatability).

Stochastic errors are characterized by a mean and a variance. The variance, as well known, can be reduced by averaging multiple measurements. The mean can be incorporated with systematic errors. The calibration of a measurement instrument is usually an off-line procedure aimed to reduce systematic errors (and indirectly affecting the mean of stochastic errors and their variance when multiple measurements are averaged). More precisely:

Statement 4.1. The calibration of an instrument measuring a quantity a is the estimation of the relationship between the measured quantity $\hat{a}$ and the actual quantity $a^* \in [a_{min}, a_{max}]$ (also called *ground-truth quantity* in the calibration procedure).

The relationship between $\hat{a}$ and a^* can be derived by two approaches:

- By means of a parametric function: $a^* = f(\hat{a}, \theta)$, where θ is a set of parameters and $f(\cdot)$ a suitable function relating $\hat{a}$ and θ to a^*. In this case the calibration becomes a parameter estimation problem and the optimal parameters value $\hat{\theta}$ can be obtained from a set of ground-truth and measured quantities $(a_1^*, \hat{a}_1), (a_2^*, \hat{a}_2), ..., (a_N^*, \hat{a}_N)$. For instance, it is rather common to estimate the calibration parameters $\hat{\theta}$ by a Mean Squared Error (MSE) approach:

$$\hat{\theta} = \underset{\theta}{\operatorname{argmin}} \frac{1}{N} \sum_{i=1}^{N} [a_i^* - f(\hat{a}_i, \theta)]^2 \tag{4.1}$$

  Since in this case the calibration is computed as a parameter estimation problem, it is addressed as *parametric calibration* or *model-based calibration*.
- By a "brute force" approach. Since the set of possible attribute values is well known ($a^* \in [a_{min}, a_{max}]$), one may sample the attribute interval obtaining N ground truth sampled values $a_1 = a_1^*, a_2 = a_2^*, .., a_N = a_N^*$. One can perform a measurement with respect to each ground truth value a_i^*, $i = 1, 2, ..., N$, obtaining therefore a set of relative measurements $\hat{a}_1, \hat{a}_2, .., \hat{a}_N$ and establish a relationship between ground-truth $a_1^*, a_2^*, .., a_N^*$ and measurements $\hat{a}_1, \hat{a}_2, .., \hat{a}_N$, by a table with N couples $(a_1^*, \hat{a}_1), (a_2^*, \hat{a}_2), ..., (a_N^*, \hat{a}_N)$ as entries. Of course, the higher is the number of interval samples (N), the more accurate is the estimated relationship but the more laborsome the calibration process. This calibration approach is usually called *non-parametric calibration* or *model-less calibration*.

The two calibration methods above have each one advantages and disadvantages. On one hand, model-based calibration needs a model which may not be always available and which may be hard to obtain. On the other hand, modeling the measurement process generally has the advantage of reducing the number of parameters to be estimated to a small set (e.g., 8 parameters for the case of a standard camera calibration), thus reducing the number N of required calibration measurements with respect to the model-less calibration case. The availability of an analytical correspondence between each possible ground-truth attribute a^* and its relative measurement $\hat{a}$ is another advantage.

Model-less calibration clearly does not need any model, but it requires a large number N of calibration measurements in order to reduce the sampling effects. This approach is more prone to over-fitting than the previous one, because N is generally way greater than the number of parameters to be estimated. There is not an analytical correspondence between each possible ground-truth value a^* and its measured value $\hat{a}$. Moreover a direct although not analytical correspondence is available only for the set of sampled ground truth values. For the other values the correspondence can be approximated by interpolation with inevitable sampling artifacts.

4.1.1 Calibration error

Calibration is not a completely precise operation, in the sense that once calibration is performed there will remain an error between estimated and actual values, commonly called *calibration error*. An estimate of such calibration error is very important since it may become the bottleneck for accuracy and precision of specific applications.

The calibration error can be computed in different ways for the case of model-based and model-less calibration upon a set of M measurements $\hat{a}_j$, $j = 1, 2, ..., M$ with relative ground-truth values a_j^*, $j = 1, 2, .., M$. Note that this set of measurements is different from the set of N measurements adopted in the calibration minimization, in order to adopt a rigorous cross-validation approach. In the case of model-based calibration, calibration error e can be computed from the M error estimates as

$$e = \frac{1}{M} \sum_{j=1}^{M} |a_j^* - f(\hat{a}_j, \hat{\theta})| \tag{4.2}$$

In the case of model-less calibration, there is no model to use in (4.2). However, given a measured quantity $\hat{a}_j$, its corresponding ground-truth value a_j^* can be approximated by interpolation: $\mathscr{F}(\hat{a}_j, (a_1^*, \hat{a}_1), .., (a_N^*, \hat{a}_N))$ where $\mathscr{F}(\cdot)$ is a suitable interpolation method (e.g., nearest-neighbor, bilinear, bicubic or splines interpolation). Calibration error e in the case of model-less calibration can be computed from the M error estimates as

$$e = \frac{1}{M} \sum_{j=1}^{M} |a_j^* - \mathscr{F}(\hat{a}_j, (a_1^*, \hat{a}_1), .., (a_N^*, \hat{a}_N))| \tag{4.3}$$

In both the cases, the goal of a calibration method is to provide a consistent procedure (several calibrations of the same device should provide similar results) giving the lowest possible calibration error.

4.1.2 Error compensation

Once the relationship between measured and actual attribute values is known, one may define a way to map the measured values to the corresponding actual values in order to compensate for the systematic differences between such quantities.

The compensation may increase the accuracy but not the precision (repeatability) of the instrument measurements, since it does not consider the error randomness. In the calibration process, and also in the actual measurement process when feasible, the precision of the device measurements can be improved by averaging over multiple measurements. However when the measurements are not performed for calibration purposes, specially in the case of dynamic scenes, averaging may not be used to reduce stochastic errors. As a simple example, consider the case of a video-camera that may only operate at $30\,[fps]$ used in a specific application requiring $30\,[fps]$, clearly in this case multiple frames cannot be averaged over time.

4.2 Calibration of color and depth imaging systems

In the specific case of computer vision or computer graphics, the attributes of interest typically are the color and/or the geometry information of a scene. The remainder of this section covers the calibration of the imaging instruments of typical computer vision and computer graphics interest, i.e., standard cameras, ToF cameras, Kinect^{TM} and stereo vision systems. In this connection the Kinect^{TM} range camera and stereo vision systems are considered as single instruments, in spite they are actually made by pairs of different devices, which in the case of stereo are two standard cameras and in the case of the Kinect^{TM} range camera are an IR projector and an IR camera. Although each camera of a stereo system individually considered is a color imaging system, in the stereo set-up they just become components of a single measurement system providing data (depth) of nature completely different with respect to that of the data acquired by its two cameras (color). It is reasonable to consider the Kinect^{TM} range camera as a single instrument since the IR projector only irradiates the scene in order to allow depth measurements and only the IR camera acts as an imaging system.

4.2.1 Calibration of a standard camera

First of all, let us recall that "standard camera" in this book refers to a digital photo-camera or to a digital video-camera, since image formation is the same for both these instruments.

The attributes measured by a standard camera are the attributes measured by all its sensor pixels. Under the hypothesis of infinitesimal size pixels, each pixel measures the color of the scene point associated to it by the perspective projection model of Section 1.2.1. Consequently, the attribute measured by a standard camera is the color set of the scene points associated to its sensor pixels.

Since the camera operation is characterized by two different processes, namely the association of the standard camera sensor pixels with the corresponding scene points and the measurement of the colors of these scene points, its calibration is divided in two parts:

- geometric calibration, i.e., the estimate of the relationship between sensor pixels positions and relative scene points;
- photometric calibration, i.e., the estimate of the relationship between the actual scene point color and the color measured by the camera at the corresponding pixel.

A comprehensive treatment of photometric calibration can be found in [1, 2]. Photometric calibration is not considered in this chapter, since for the applications treated in this book only geometrical calibration is relevant.

Statement 4.2. The geometric calibration of a video camera is the estimation of the relationship concerning measured and actual correspondences between camera sensor pixels and 3D scene points.

In order to clarify the above statement, let us recall that according to the pin-hole camera model introduced in Section 1.2.1, the image of a standard camera is obtained by perspective projection, i.e., as intersection on the sensor plane of all the rays connecting the scene points P with O (corresponding to the position of the nodal point of the optics).

As pictorially shown in Figure 4.1, the same 3D scene point P is always projected to the same camera pixel p as long as the camera stays still, but any point of the optical ray (green line) is associated to the same pixel p (green point on the sensor). From the set of all the scene points projected to p, the scene point P actually projected to p is the one with the smallest radial distance from p.

The geometrical calibration of a standard camera can be regarded as the estimation of the relationship between the actual and the measured optical rays relative to each camera sensor pixel. In case of ideal (i.e., distortion-free) optics the optical ray is determined by the camera intrinsic parameters only by (1.6). In the case of real optics characterized by radial and tangential distortion the geometry of the optical rays is determined by both the camera intrinsics and the optics distortion parameters. As already seen, the directions of the optical rays can be modeled by the Heikkila model (1.12) [3] or by more complex models such as the fractional model [4].

Rephrasing the above according to the notation of Section 4.1, the attribute a of geometrical camera calibration is the geometry (origin and orientation) of the optical ray. Its measured value $\hat{a}$ in the case of the Heikkila camera model[1] is given by

---

[1] For conciseness sake, this is the only camera distortion model considered in this book, however all the presented consideration can be similarly applied to the case of the fractional camera model.

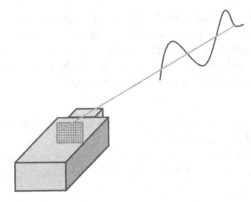

Fig. 4.1 Non-univocal association of a single camera sensor pixel (green dot) to 3D scene points (green line).

all the parameters of the camera projection matrix (1.11) plus the distortion parameters, namely by the intrinsic parameters of matrix (1.7), i.e., $\hat{f}_x$, $\hat{f}_y$, $\hat{c}_x$ and $\hat{c}_y$ by the estimated radial and tangential distortion parameters $\hat{\mathbf{d}} = [\hat{k}_1, \hat{k}_2, \hat{k}_3, \hat{d}_1, \hat{d}_2]$ of (1.12) and in general by the extrinsic parameters $\widehat{R}$ and $\hat{\mathbf{t}}$ of (1.9) relating the optical ray geometry with respect to the CCS to the WCS. The relationship between the camera parameters, measured quantities and actual values $a^* = f(\hat{a}, \hat{\theta})$ is given by (1.11) and (1.12).

A detailed analysis of calibration parameters and relationships is beyond the scope of this book and can be found in classical computer vision readings, such as [3, 4, 5, 6, 7].

Any calibration technique adopts calibration objects of known size, offering highly identifiable saliency points to be used as ground truth values a^*.

A popular calibration object [3, 5] is a black and white checkerboard, with checkers of known size. In this case the saliency points serving as a^* are the the checkerboard corners (that can be easily detected), as shown by Figure 4.2.

The WCS can be defined as the reference system with origin at the checkerboard top left corner, x and y axis on the checkerboard plane and z axis orthogonal to the checkerboard plane as shown in Figure 4.3.

Due to the planar shape of the sensor and checkerboard, the position of the sensor plane with respect to the WCS, associated to the checkerboard, can be derived by a 3D homography [5,4], characterized by 9 coefficients. Once the 3D position of the sensor with respect to the checkerboard is known, as shown by Figure 4.3, the 3D coordinates of the checkerboard corners, P_i^*, serving as ground truth, can be projected to the sensor by (1.10) giving ground-truth pixel values p_i^*.

The coordinates of the ground-truth pixels p_i^* can be compared with those of the pixels $\hat{p}_i$, directly detected from the measured checkerboard image, obtaining an error expression which depends on projection matrix (1.11) and distortion parameters (1.12). The sum of such errors, corresponding to the M different spatial checker-

Fig. 4.2 Corners detection on a calibration checkerboard.

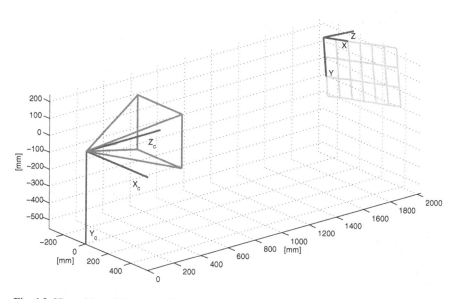

Fig. 4.3 3D position of the sensor plane with respect to the checkerboard.

board positions, gives an expression of the model-based calibration error of the type
of Equation (4.2), which is function of camera calibration parameters (1.11) and
(1.12). This expression can be used to determine the intrinsic camera parameters,
together with the extrinsic parameters $(R_k, \mathbf{t_k})$ with $k = 1, 2, ..., M$ giving the camera
sensor position with respect to each spatial checkerboard position. This procedure,
due to [5], is the standard method for the estimate of the Heikkila model parameters, also available in popular open source camera calibration projects, such as the
MATLAB Camera Calibration Toolbox [8] and OpenCV [9].

4.2.1.1 Standard camera systematic error compensation: undistortion

In the case of standard cameras, lens distortion causes a systematic error of the type
indicated in Section 4.1.2. Once the standard camera is calibrated, its distortion
parameters are known and the image undistortion procedure (1.12) can be applied
to all the images acquired by the standard camera in order to compensate for its
radial and tangential distortion, i.e., for its systematic error in the measure of the
optical rays orientations.

4.2.2 Calibration of a ToF camera

Since ToF camera image formation can be modeled by perspective projection, ToF
camera calibration is strongly reminiscent of standard camera calibration. As in that
case, indeed, the attributes measured by a ToF camera are the attributes measured
by the ToF camera sensor pixels. Under the hypothesis of infinitesimal size pixels,
each pixel measures the radial distance from the scene points associated to it by the
perspective projection model of Section 1.2.1. Therefore, the attribute measured by
a ToF camera is the radial distance of the points associated to the ToF camera sensor
pixels. The ToF operation is characterized by two different processes:

- the association of the ToF camera sensor pixels to the relative scene points;
- the measurements of the radial distances of the scene points.

ToF camera calibration reflects this measurements sub-division according to [10],
and, with the terminology introduced for standard camera calibration, it can be divided in:

- geometrical calibration, i.e., the estimate of the relationship between sensor pixels positions and relative scene points;
- photometric calibration, i.e., the estimate of the relationship between the actual
radial distance of the scene point and the radial distance measured by the ToF
camera at the corresponding pixel.

The geometrical calibration of a ToF camera is identical to the one of a standard
camera and it can be made by the techniques indicated in the previous sub-section

with the peculiarity that ToF camera geometrical calibration must use amplitude images A_T instead of standard color (or gray-level) images as shown by Figure 4.4.

Fig. 4.4 Example of the IR image of a checkerboard acquired by a ToF camera with the corners detected.

The amplitude images can be collected in two different ways, either in the so-called *standard mode*, i.e., with the ToF camera illuminators active during acquisition, or in the so-called *common-mode*, i.e., with ToF camera illuminators off during acquisition, namely using the ToF camera as a standard IR camera. The first solution is more direct and generally produces better results, but it requires proper integration time setting in order to avoid saturation. The second solution requires an external auxiliary IR illumination system.

The photometric ToF camera calibration deals with the so-called *systematic offsets* of ToF camera radial distance measurements, due to various factors, such as harmonic distortion and non-zero mean Poisson distributed photon-shot noise (as explained in Chapter 2). It is worth recalling that harmonic distortion depends on the distance from the object to be measured, and that the photon-shot noise depends on the received signal intensity and amplitude. Therefore the systematic offsets ultimately depend on the measured distance, amplitude and intensity. The calibration of systematic offsets can be performed by measuring targets characterized by different reflectivity placed at known distances and by comparing the measured distances against the actual ones. A comprehensive description and analysis of such procedures can be found in [10, 11, 12]. Such measurements can be used both in a model-based approach as in [11] where the model is a polynomial function, or in a model-less approach as in [10] where a table describes the relationship between measured and actual radial distances.

4.2.2.1 Compensation of ToF cameras systematic errors

The compensation of the systematic ToF cameras errors includes both the geometrical calibration errors (i.e., undistortion) and the radial distance measurements errors.

The two compensations can be performed independently, like the relative calibrations. Undistortion can be implemented as for the case of standard cameras. Systematic radial distance measurements errors in the case of model-based calibration can be compensated by inverting the fitted functional-model and in the case of model-less calibration by a look-up-table correction.

4.2.3 Calibration of the Kinect^{TM} range camera

As in the previous cases, the attributes measured by the Kinect^{TM} range camera are the attributes measured by its IR camera sensor pixels. Under the hypothesis of infinitesimal size pixels, each pixel measures the disparity of the associated scene point with respect to a reference image. Therefore, the attributes measured by the Kinect^{TM} range camera are the disparities of the scene points associated to its IR camera sensor pixels. The set of acquired disparity values defines the disparity map of the scene.

Since the Kinect^{TM} range camera is made by an IR camera and an IR projector, its calibration in principle could be performed by the method of [13], which also provides a way to compensate for the systematic measurements errors.

Figure 4.5 shows the IR image of a checkerboard taken by the Kinect^{TM} range camera with the corners detected. This kind of images can be used to calibrate the Kinect^{TM} if its calibration parameters are not accessible by the users and it can prove useful when the Kinect^{TM} is used together with a standard camera, as it will be shown in Section 4.3.1.

Fig. 4.5 Example of the IR image of a checkerboard acquired by a Kinect^{TM} range camera.

The actual procedures adopted by Primesense in order to calibrate and compensate the Kinect^{TM} range camera measurements have not been disclosed yet. This omission however does not penalize the usage of Kinect^{TM} range cameras, since they are supplied already calibrated and with built-in compensation.

4.2.4 Calibration of a stereo vision system

The attribute measured by a stereo vision system S is the depth of the scene points framed by both of its cameras. Let us recall that the 3D coordinates $\mathbf{P} = [x, y, z]^T$ of P with respect to the S reference system can be estimated by triangulation from the 2D coordinates of conjugate points p_L and p_R via (1.14) and (1.15). The triangulation procedure depends on the properties of the two standard cameras L and R and on their relative position. The properties of L and R are parametrically modeled as seen in Section 4.2.1 by their intrinsic and distortion parameters, while their relative position is parametrically modeled by the relative roto-translation between the L-3D and the R-3D reference systems. Therefore stereo vision system calibration is customarily approached as a parametric model-based calibration problem.

Statement 4.3. The calibration of a stereo vision system (or stereo calibration) is the estimation of the following quantities:

- the intrinsic parameters matrices K_L and K_R
- the distortion parameters (such as $\mathbf{d_L}$ and $\mathbf{d_R}$ defined by (1.12)) of the L and R cameras respectively
- the 3×3 rotation matrix R and the 3×1 translation vector $\mathbf{t}$ describing the roto-translation between the L-3D and the R-3D reference systems.

Clearly the two intrinsic parameters matrices K_L and K_R and the distortion parameters can be derived independently by single L and R camera calibrations. Therefore, common stereo calibration procedures usually perform first the two standard camera calibrations in order to obtain the intrinsic parameters, and then estimate the stereo system extrinsic parameters R and $\mathbf{t}$. Such a two-step operation allows to reduce the size of the estimated parameters space, greatly simplifying the estimation problem. A comprehensive treatment of stereo calibration can be found in [6]. Popular open-source stereo calibration are in the MATLAB Camera Calibration Toolbox [8] and in the OpenCV Library [9]. Figure 4.6 shows the relative positions of the L and R camera sensors with respect to the calibration checkerboards.

4.2.4.1 Compensation and Rectification

Stereo images rectification is a procedure transforming the images acquired by a stereo vision system in such a way that conjugate points (p_L and p_R) share the same vertical coordinate ($v_L = v_R$) in the L-2D and in the R-2D reference systems respectively. Stereo images rectification is often associated to stereo calibration since stereo images rectification simplifies and makes more robust 3D depth estimation by stereo vision algorithms. Once a stereo vision system is calibrated, it is always possible to apply a stereo image rectification procedure [2]. A popular stereo images rectification algorithm is given in [14].

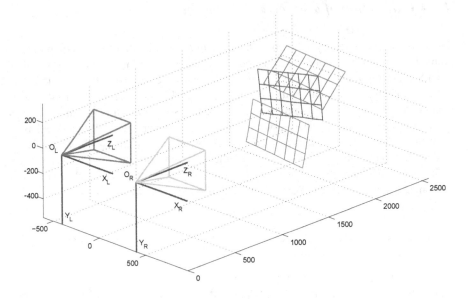

Fig. 4.6 3D position of the L and R camera sensors with respect to the checkerboards.

The images acquired by the two cameras of a stereo vision system are typically affected by radial and tangential distortion. Undistortion and rectification can be applied in a single step. The possibility of combining the two procedures may hide the fact that only the former is about systematic errors compensation but not the latter. Namely, the undistortion of the images acquired by L and R is actually a compensation of systematic errors, but rectification does not compensate for systematic errors, since it just transforms the acquired images in order to simplify the tasks of stereo vision algorithms.

4.3 Calibration of heterogeneous imaging systems

An heterogeneous measurement system is a set-up made by multiple measurement instruments of different nature measuring the attributes of the same object. Heterogeneous acquisition systems can be subdivided into two different groups, i.e., systems made by devices measuring the same attribute and systems made by devices measuring different attributes.

This section considers only instruments measuring visual quantities, i.e., *imaging systems* measuring color and geometry. The visual attribute measured by a specific imaging instrument depends on its spatial position, hence the measurements of different imaging systems can only be related upon the knowledge of their relative positions. These considerations can be summarized as follows.

Statement 4.4. The calibration of an heterogeneous imaging system is the procedure made by the calibration of each imaging system forming the heterogeneous
system and by the estimate of their relative positions.

The remainder of this section considers the calibration of a system made by a ToF
camera and a color camera, of a system made by a Kinect^{TM} range camera and
a color camera (which are both examples of imaging systems measuring different
visual attributes such as geometry and color) and of a system made by a ToF camera
and a stereo vision system (which is an example of imaging systems measuring the
same visual attribute, i.e., the scene geometry).

4.3.1 Calibration of a system made by a standard camera and a ToF camera or a Kinect^{TM}

Let us consider first the calibration of a system made by a standard camera and a ToF
camera or a Kinect^{TM}, i.e., an heterogeneous system supporting an imaging system
measuring the scene color and an imaging system measuring the scene geometry
(the ToF camera or the Kinect^{TM} range camera).

The first step is the independent calibration of all the instruments forming the system. In the case of the first system, made by a ToF camera and a standard camera,
it is therefore necessary to calibrate and compensate the measurements of the ToF
camera and to calibrate and compensate the measurements of the standard camera.
In the case of the second system, made by a Kinect^{TM} range camera and a standard
camera, it is therefore necessary to calibrate and compensate the measurements of
the Kinect^{TM} range camera and to calibrate and compensate the measurements of
the standard camera. Once the ToF camera or the Kinect^{TM} range camera are calibrated, they can be considered functionally equivalent, since they both perform the
same kind of measurements, i.e., they provide depth or 3D geometry measurements.

The second step is the estimation of the relative positions of the two instruments
forming the heterogeneous system, i.e., the estimate of the relative roto-translation
between the reference systems associated to the two instruments. This can be ac-
complished by the recognition of saliency points of suitable calibration objects (e.g.,
checkerboards) by the different instruments. For instance, it is possible to identify
the corners of a (black and white) checkerboard both on the images I_C acquired by
the standard cameras and on the IR images A_T acquired by a ToF camera or on the
IR images I_K acquired by a Kinect^{TM} range camera, as exemplified in Figures 4.4
and 4.5.

From the 2D coordinates $\mathbf{p^i} = [u^i, v^i]^T$ of a point p^i in a undistorted image acquired
by a calibrated camera (either C, T or K), obtained as a corner of a calibration
checkerboard, one may compute the 3D coordinates $\mathbf{P^i} = [x^i, y^i, z^i]^T$ of the corre-

sponding scene point P^i with respect to the 3D camera reference system and estimate the 3D homography between the sensor plane and the checkerboard plane as
depicted in Figure 4.7. This procedure is described in [5] and implemented in [8, 9].

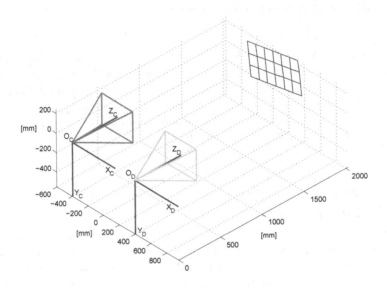

Fig. 4.7 Estimation of the 3D coordinates of the corners in [*mm*] of a checkerboard (corners of the
blue grid) from their 2D coordinates on an image acquired by a calibrated standard camera C (red
pyramid) and from an image acquired by a depth camera (green pyramid). The distance between
O_C and O_D should be as small as possible: the pictorial camera models are out of scale.

In the case of a range camera D (either a ToF camera T or a Kinect^{TM} range
camera K), the 3D coordinates of the checkerboard corners can also be directly
estimated by the depth information acquired by the depth camera. Indeed, a depth
camera D can provide both the undistorted 2D coordinates $\mathbf{p}_D^i = [u_D^i, v_D^i]^T$ of a point
p_D^i (checkerboard corner) on its IR image, here denoted as I_D for notation semplicity,
and a direct estimate of the depth z_D^i of the corresponding 3D point P_D^i. From this
datum the 3D coordinates $\mathbf{P}_D^i = [x_D^i, y_D^i, z_D^i]^T$ of P_D^i can be computed by inverting
the projection Equation (1.6), as hinted by Figure 4.8 where the 3D positions of the
checkerboard corners, directly estimated by the depth cameras, are denoted by green
dots.

Therefore for a range camera D there are two different ways of computing the 3D
coordinates of the checkerboard, namely the homography-based procedure and the
direct depth measurement. The main difference between the two approaches is that
the homography-based approach exploits the ground-truth geometrical properties of
the calibration checkerboard, while the direct depth measurements does not.

For practical purposes some details need to be underlined. In the case of a ToF

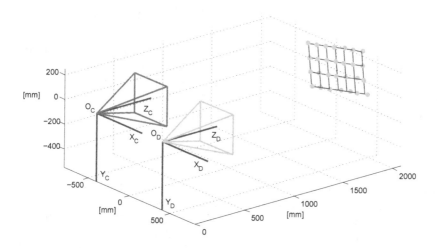

Fig. 4.8 Direct computation of the 3D coordinates of the corners of a checkerboard (green dots) by a depth camera (D) (green pyramid) and 3D position with respect to the checkerboard of a standard camera (red pyramid).

camera T it is possible to extrapolate both the 2D coordinates $\mathbf{p_D^i}$ from A_T and the depth estimate of z_D^i from Z_T in a single step. In the case of a Kinect™ range camera K, instead, this requires two steps since the acquisition of the 2D coordinates of the checkerboard corner needs to acquire IR images I_K without pattern projection and the acquisition of their 3D coordinates needs the depth map Z_K which requires the pattern projection. Summarizing, the needed information for each position of the checkerboard requires a single acquisition in the case of a ToF camera and two acquisitions in the case of a Kinect™ range camera.

The 3D coordinates $\mathbf{P_C^i} = [x_C^i, y_C^i, z_C^i]^T$ of the checkerboard corners $P_C^i, i = 1,...,N$ computed from the measurements of standard camera C with respect to the C-3D reference system, and the 3D coordinates $\mathbf{P_D^i} = [x_D^i, y_D^i, z_D^i]^T$ of the checkerboard corners $P_D^i, i = 1,...,N$ computed from the measurements of D with respect to the D-3D reference system, are related by the rototranslation $(R, \mathbf{t})$ between the C-3D and the D-3D reference systems. In other words, the standard camera C and depth camera D provide the same cloud of points expressed with respect to two different reference systems. The computation of the roto-translation between the same point-clouds in two different reference systems is known as the *absolute orientation problem* and has a closed-form solution given by the Horn Algorithm [15], which estimates the rotation matrix $\hat{R}$ and the translation vector $\hat{\mathbf{t}}$ minimizing the distance between the 3D positions of the checkerboard corners measured by the two imaging systems C and D, namely

$$\left[\hat{R}, \hat{\mathbf{t}}\right] = \underset{[R,\mathbf{t}]}{\operatorname{argmin}} \frac{1}{N} \sum_{i=1}^{N} \|\mathbf{P_C^i} - [R\mathbf{P_D^i} + \mathbf{t}]\|_2 \tag{4.4}$$

An alternative way of computing $(\hat{R}, \hat{\mathbf{t}})$ is to reproject points P_D^i on the camera sensor by the camera intrinsic matrix K_C and estimate the R and $\mathbf{t}$ which minimize the difference between the pixels p_C^i, $i = 1, 2, ..., N$ with coordinates $\mathbf{p_C^i} = [u_C^i, v_C^i]$ of the images acquired by the standard camera I_C corresponding to the checkerboard corners and the projection of P_D^i on the camera sensor image plane, as

$$\left[\hat{R}, \hat{\mathbf{t}}\right] = \underset{[R,\mathbf{t}]}{\operatorname{argmin}} \frac{1}{N} \sum_{i=1}^{N} \left\| \begin{bmatrix} \mathbf{p_C^i} \\ 1 \end{bmatrix} - K_C[R\mathbf{P_D^i} + \mathbf{t}] \right\|_2 \tag{4.5}$$

This minimization can be solved by classical optimization techniques (e.g., gradient descent) starting from the estimates of R and $\mathbf{t}$ obtained from the Horn algorithm.

As it will be shown in the subsequent chapters of this book, there are applications requiring to associate the 3D coordinates of a scene point acquired by a depth camera D with its color acquired by a color camera C. The color of the point is typically obtained by back-projecting the 3D point computed by D to the color image I_C (and by interpolating the color value in correspondence of the typically non-integer coordinates of the back-projected point).

Summarizing, there are two main algorithmic choices in the calibration of a system made by a standard camera C and a range camera D (either a ToF camera T or a Kinect^{TM} range camera K), namely:

- the first choice is the selection between the homography-based and the direct depth measurement approach for obtaining the 3D coordinates of the checkerboard corners acquired by D. The main advantage of the former approach is that the calibration precision and accuracy are not affected by the precision and the accuracy of the depth measurements of D. The main advantage of the latter approach, instead, is that the calibration accounts for the nature of the measurements performed by D, leading therefore to a better synergy between the data acquired by the two imaging systems.
- The second choice concerns the selection of the optimization criterion for the estimation of R and $\mathbf{t}$. The two options are the *absolute-orientation approach* and the *reprojection-error minimization approach*. The former generally implies an easier task and an exact solution, but the latter leads to better performances for most applications.

The calibration errors can be computed with respect to a set of 3D points P^j, $j = 1, 2, ..., M$ different from the points P^i, $i = 1, 2, ..., N$ used to compute $\hat{R}$ and $\hat{\mathbf{t}}$ in (4.4) or (4.5). In the case of $\left[\hat{R}, \hat{\mathbf{t}}\right]$ obtained from (4.4) it can be computed as:

$$e_1 = \frac{1}{J} \sum_{j=1}^{M} \|\mathbf{P_C^j} - [\hat{R}\mathbf{P_D^j} + \hat{\mathbf{t}}]\|_2 \tag{4.6}$$

Instead in case of $\left[\hat{R}, \hat{\mathbf{t}}\right]$ obtained from (4.5) it is given by:

$$e_2 = \frac{1}{J} \sum_{j=1}^{M} \left\| \begin{bmatrix} \mathbf{P_C^j} \\ 1 \end{bmatrix} - K_C \left[\mathbf{P_C^j} - [\hat{R}\mathbf{p_D^j} + \hat{t}] \right] \right\|_2 \tag{4.7}$$

Calibration error e_1 is measured in $[mm]$ and it is typically smaller than $10[mm]$. Calibration error e_2 is measured in $[pxl]$ and it is typically smaller than $1[pxl]$, with XGA camera with $3.5mm$ optics.

4.3.2 Calibration of a system made by a stereo vision system and a ToF camera or a Kinect^{TM}

The calibration of a system made by a stereo vision system S and an alternate depth measurement system D made by a ToF camera T or a Kinect^{TM} range camera K is considered next. The most common methods proposed in order to accomplish this task are the ones of [10, 16]. The heterogeneous system in this case is made by two depth cameras of different type.

As in the cases of Section 4.3.1, the first step of the heterogeneous systems calibration is the independent calibration of the depth camera D and of the stereo vision system S. The two standard cameras of S must also be rectified, and the images I_S acquired by S are assumed to be compensated (i.e., undistorted) and rectified.

The second step of the heterogeneous systems calibration is the estimation of the relative positions of S and D, i.e., the estimation of the roto-translation $(R, \mathbf{t})$ between the S-3D and the D-3D reference systems. The procedure for estimating $(R, \mathbf{t})$ can be summarized as:

1. Extraction of the calibration checkerboard corners from the color images I_S. This gives a set of pixels p_L^i, $i = 1, ..., N$ with coordinates $\mathbf{p_L^i} = [u_L^i, v_L^i]^T$ on the images acquired by the left camera L of the stereo pair and the conjugate pixels p_R^i, $i = 1, .., N$ with coordinates $\mathbf{p_R^i} = [u_L^i - d, v_L^i]^T$ on the images acquired by the right camera R, where d is the *disparity*.
2. Extraction of the checkerboard corners from the IR images I_D acquired by D. This gives a set of pixels p_D^i, $i = 1, 2, ..., N$ on the images I_D.
3. Computation of the 3D coordinates of the two sets of corners with respect to the S-3D and D-3D reference systems respectively. The stereo system S computes by triangulation 3D points P_S^i, $i = 1, 2, ..., N$ with coordinates $\mathbf{P_S^i} = [x_S^i, y_S^i, z_S^i]^T$, with respect to the S-3D reference system relative to the pairs (p_L^i, p_R^i), $i = 1, .., N$.
The depth camera D associates the pixels p_D^i, $i = 1, 2, ..., N$ of the IR images to the points P_D^i, $i = 1, 2, ..., N$ directly computed by the depth-camera, which have coordinates $\mathbf{P_D^i} = [x_D^i, y_D^i, z_D^i]^T$ with respect to the D-3D reference system.
4. Estimation of the roto-translations $(R, \mathbf{t})$ between the 3D reference systems of the two depth cameras either by solving the absolute orientation problem or by minimizing the reprojection error.

The estimation of $(R, \mathbf{t})$ by Horn algorithm in this case can be written as

$$[\hat{R}, \hat{\mathbf{t}}] = \underset{[R,\mathbf{t}]}{\mathrm{argmin}} \frac{1}{N} \sum_{i=1}^{N} \left\| \mathbf{P_S^i} - \left[R\mathbf{P_D^i} + \mathbf{t} \right] \right\|_2 \tag{4.8}$$

The estimation of $(R, \mathbf{t})$ can also be performed similarly to (4.5) by accounting for the reprojection error on the L and R images, as

$$[\hat{R}, \hat{\mathbf{t}}] = \underset{[R,\mathbf{t}]}{\mathrm{argmin}} \frac{1}{2N} \sum_{i=1}^{N} \left[\left\| \begin{bmatrix} \mathbf{p_L^i} \\ 1 \end{bmatrix} - K_S \left[R\mathbf{P_D^i} + \mathbf{t} \right] \right\|_2 + \left\| \begin{bmatrix} \mathbf{p_R^i} \\ 1 \end{bmatrix} - K_S \left[R\mathbf{P_D^i} + \mathbf{t} - \mathbf{b} \right] \right\|_2 \right] \tag{4.9}$$

where K_S is the intrinsic parameters matrix of both the L and R cameras of the rectified stereo vision system with baseline b and $\mathbf{b} = [-b, 0, 0]^T$ is the vector with the coordinates of the origin of the R-3D reference system with respect to the L-3D reference system. The expression of $\mathbf{b}$ assumes that the L-3D and R-3D reference systems have the axis oriented according to the right-hand orientation convention as shown in Figure 1.6 (we assumed that the L-3D reference system is the S-3D reference system).

In the case of an heterogeneous system made by a range camera and a stereo vision system, the choice between absolute orientation or reprojection error minimization depends on the specific application. For instance, if the application calls for the fusion of the 3D scene geometry estimates of the stereo vision system and of the depth camera, it is worth optimizing absolute orientation, since in this case a good synergy between the data acquired by the two depth imaging systems is important. If the application, instead, reprojects the 3D geometry estimates performed by D on the images acquired by L and R for further processing, it is worth minimizing the reprojection error.

The calibration error can be computed with respect to a set of points P_S^j and P_D^j with $j = 1, 2, ..., M$ different from the points used to estimate R and $\mathbf{t}$ in (4.8) and (4.9). In case of $[\hat{R}, \hat{\mathbf{t}}]$ given by (4.8) the calibration error is

$$e_1 = \frac{1}{J} \sum_{j=1}^{M} \left\| \mathbf{P_S^j} - [\hat{R}\mathbf{P_D^j} + \hat{\mathbf{t}}] \right\|_2 \tag{4.10}$$

and it is measured in $[mm]$, while in case of $[\hat{R}, \hat{\mathbf{t}}]$ given by (4.9) it is

$$e_2 = \frac{1}{2J} \sum_{j=1}^{M} \left[\left\| \begin{bmatrix} \mathbf{p_L^j} \\ 1 \end{bmatrix} - K_S [\hat{R}\mathbf{P_D^j} + \hat{\mathbf{t}}] \right\|_2 + \left\| \begin{bmatrix} \mathbf{p_R^j} \\ 1 \end{bmatrix} - K_S [\hat{R}\mathbf{P_D^j} + \hat{\mathbf{t}} - \mathbf{b}] \right\|_2 \right] \tag{4.11}$$

and it is measured in $[pxl]$.
Calibration error e_1 is typically smaller than $10[mm]$ and e_2 is typically smaller than $1[pxl]$, with XGA cameras with $3.5[mm]$ optics.

4.4 Conclusions and further readings

The ultimate goal of this chapter is to give the notions needed to jointly calibrate standard cameras and depth cameras, such as a ToF camera and a Kinect^{TM} range camera, for their combined usage. This requires to consider first the calibration of individual color and depth imaging systems.

The calibration of a standard camera is of fundamental relevance for the calibration of all the other imaging systems, both from the conceptual and the operational point of view. The reader interested to more details about camera calibration can find image processing and corner detection techniques in [17, 18], projective geometry in [7, 19] and numerical methods in [20]. Classical camera calibration approaches are the ones of [3, 4, 5]. Some theoretical and practical hints can also be found in [6].

The principles of stereo vision system calibration are reported in [6], and open source implementations of stereo vision calibration algorithms are in the MATLAB Camera Calibration Toolbox [8] and in the calibration routines of OpenCV [9].

ToF camera calibration theory and practice is presented in [10, 12, 11]. A basic fundamental reading for the calibration of matricial active triangulation cameras, such as the Kinect^{TM} range camera, is [13]. An example of calibration of a system made by a standard camera and a range camera is reported in [21] and the calibration of heterogeneous systems made by a stereo vision system and by a range camera is reported in [10, 16, 11].

References

1. C. Steger, M. Ulrich, and C. Wiedemann, *Machine Vision Algorithms and Applications*. Wiley-VCH, 2007.
2. R. Szeliski, *Computer Vision: Algorithms and Applications*. New York: Springer, 2010.
3. J. Heikkila and O. Silven, "A four-step camera calibration procedure with implicit image correction," in *CVPR*, 1997.
4. D. Claus and A. Fitzgibbon, "A rational function lens distortion model for general cameras," in *CVPR*, 2005.
5. Z. Zhang, "A flexible new technique for camera calibration," *IEEE Transactions on Pattern Analysis and Machine Intelligence*, vol. 22, pp. 1330–1334, 1998.
6. G. Bradski and A. Kaehler, *Learning OpenCV: Computer Vision with the OpenCV Library*. O'Reilly, 2008.
7. R. Hartley and A. Zisserman, *Multiple View Geometry in Computer Vision*. Cambridge University Press, 2004.
8. J.-Y. Bouguet, "Camera calibration toolbox for matlab."
9. "OpenCV." http://opencv.willowgarage.com/wiki/.
10. J. Zhu, L. Wang, R. Yang, and J. Davis, "Fusion of time-of-flight depth and stereo for high accuracy depth maps," in *CVPR*, 2008.
11. I. Schiller, C. Beder, and R. Koch, "Calibration of a pmd-camera using a planar calibration pattern together with a multi-camera setup," in *Proc.of ISPRS Conf.*, 2008.
12. T. Kahlmann and H. Ingensand, "Calibration and development for increased accuracy of 3d range imaging cameras," *Journal of Applied Geodesy*, vol. 2, pp. 1–11, 2008.

13. M. Trobina, "Error model of a coded-light range sensor," tech. rep., Communication Technology Laboratory Image Science Group, ETH-Zentrum, Zurich, 1995.

14. A. Fusiello, E. Trucco, and A. Verri, "A compact algorithm for rectification of stereo pairs," *Machine Vision and Applications*, vol. 12, pp. 16–22, 2000.

15. B. Horn, "Closed-form solution of absolute orientation using unit quaternions," *Journal of the Optical Society of America*, vol. 4, pp. 629–642, 1987.

16. C. Dal Mutto, P. Zanuttigh, and G. Cortelazzo, "A probabilistic approach to tof and stereo data fusion," in *Proceedings of 3DPVT*, (Paris, France), May 2010.

17. R. Gonzalez and R. Woods, *Digital Image Processing*. Addison-Wesley Longman Publishing Co., Inc., 2001.

18. J. Shi and C. Tomasi, "Good Features to Track," in *CVPR*, 1994.

19. E. Trucco and A. Verri, *Introductory Techniques for 3-D Computer Vision*. Prentice Hall PTR, 1998.

20. J. Nocedal and S. Wright, *Numerical Optimization*. Springer, 2000.

21. V. Garro, C. Dal Mutto, P. Zanuttigh, and G. Cortelazzo, "A novel interpolation scheme for range data with side information," in *CVMP*, pp. 52 –60, nov. 2009.

Chapter 5
Fusion of Depth Data with Standard Cameras Data

Depth camera possibilities and limitations seen in the previous chapters may naturally prompt questions like "Is a depth camera enough for my application?" or "May one or more standard cameras help it?".

Depth cameras can only provide 3D geometry information about the scene. If an application needs color, the usage of one or more color standard cameras becomes necessary. This is for instance the case of scene segmentation by means of color and geometry (treated in Chapter 6).

ToF cameras are generally characterized by low spatial resolution (Chapter 2) and Kinect^{TM} by poor edge localization (Chapter 3). Hence a depth camera alone is not suited for the estimate of precise high-resolution 3D geometry near depth discontinuities. If such information is desired, it is worth coupling a depth camera with a standard camera. For instance, hand or body gesture recognition and alpha-matting applications might benefit from this kind of heterogeneous acquisition systems.

If the accuracy or the robustness of depth cameras measurements are not adequate, in particular in the near range, one may consider an acquisition system made by a depth camera and a stereo system. This solution can also reduce occlusions effects between color and 3D geometry information. 3D video production and 3D reconstruction are applications where this kind of setup may be beneficial.

This chapter analyzes all the issues related to the combination of depth data with standard cameras. For a synergic data combination it is firstly necessary to register the depth and standard camera data as discussed in the next section. The usage of a standard camera together with a depth camera allows to increase the spatial resolution of the depth map. This is why this possibility is called spatial super-resolution. Super-resolution can be obtained either by deterministic or probabilistic methods, each one with advantages and disadvantages, as explained in Section 5.2 and 5.3. Another intriguing possibility, treated in Section 5.4 is the fusion of the depth data obtained from a depth camera with the depth data obtained by a stereo system.

5.1 Acquisition setup and data registration

Figure 5.1 shows some examples of the two types of acquisition systems considered in this chapter. The first one is made by a depth camera (either a Kinect^{TM} or a ToF) and a single standard camera, while the second one by a depth camera and a couple of standard cameras.

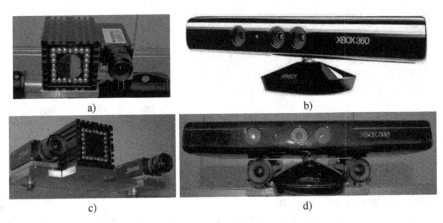

Fig. 5.1 Acquisition setup made by a) a MESA Imaging SR4000 and one color camera; b) the Kinect^{TM} supports an internal color camera; c) a MESA Imaging SR4000 and two color cameras used as an autonomous stereo vision system; d) a Kinect^{TM} and two color cameras used as an autonomous stereo vision system.

Let us consider first the setup made by a ToF camera T and a standard color camera C. Note that if C and T are in different positions the color and depth streams refer to two different viewpoints. This can only be avoided by using an optical splitter on both devices as proposed in [1]. This approach, however, increases system costs and introduces a number of undesirable effects. Notably the splitter affects the optical power and thus the distances measured by the ToF [19] and the IR emitters must be moved out of the ToF camera. For all these reasons, it is way simpler to place a ToF and a standard camera on a rig as close as possible.

In the case of the Kinect^{TM} both its depth and color cameras are embedded in it, but even though camera and depth sensor are placed very close, they are not co-positioned.

Either in the case of a color camera coupled with a ToF camera or in the case of the Kinect^{TM}, the output data will be of the same type, i.e., a color and a depth stream relative to two slightly different viewpoints.

Figure 5.2 (a-d) shows an example of the data acquired by a color camera (i.e., an high resolution color image a)) combined with a ToF camera (i.e., a low resolution depth map b), a low resolution intensity image c) and a low resolution confidence map d)). Figure 5.2 (e-g) shows the data of the same scene acquired by the Kinect^{TM}.

Figure 5.3 shows instead an example of how, no matter how closely a standard camera and a ToF camera are placed, their two images will refer to slightly different viewpoints, and some regions visible from the camera viewpoint will not be visible from the ToF viewpoint or vice-versa.

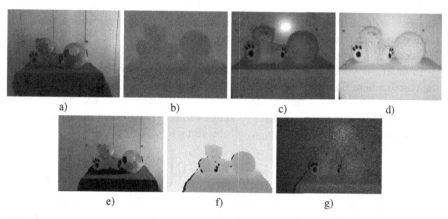

a)   b)   c)   d)

e)   f)   g)

Fig. 5.2 Example of data acquired by a setup made by a ToF camera and a standard camera (a-d) and by a Kinect™ (e-g): a) color image acquired by the standard camera; b) depth map acquired by the ToF camera ; c) amplitude image acquired by the ToF camera; d) confidence map of the ToF camera; e) color image acquired by the Kinect; f) depth map acquired by the Kinect™; g) amplitude image acquired by the Kinect™.

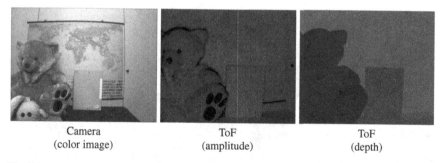

Camera   ToF   ToF
(color image)   (amplitude) (depth)

Fig. 5.3 Example of occlusion issues: note how the background in the region between the teddy-bear and the collector is visible from the camera viewpoint but not from the viewpoint of the ToF camera.

Let us recall that it is necessary to calibrate the acquisition set-up as seen in Chapter 4 in order to obtain a common reference system for the color and the depth data streams provided by the two imaging systems. Once a common reference system is available, one may obtain the following types of output:

- a color image and a depth map corresponding to the viewpoint of the color camera;
- a color image and a depth map corresponding to the viewpoint of the ToF camera (or of the Kinect^{TM} IR camera);
- a colored 3D point cloud representing the scene;
- a textured 3D mesh representing the scene.

Images and depth maps are typically more suited to 3D video applications, while 3D textured meshes or colored point clouds are more suited to applications concerning 3D scene reconstruction or navigation.

In order to obtain an image I_C and a depth map Z_C referred to the viewpoint of the color camera C defined on lattice Λ_C, each sample $p_D \in Z_C$ which is associated to a 3D point P_D with coordinates $\mathbf{P}_D = [x_D, y_D, z_D]^T$ acquired by the range camera D (either a ToF camera or a Kinect^{TM} range camera) must be reprojected to pixel $p_C \in I_C$ with coordinates $\mathbf{p}_C = [u_C, v_C]^T$ according to:

$$\begin{bmatrix} \mathbf{p}_C \\ 1 \end{bmatrix} = \frac{1}{z_C} \, \hat{K}_C \begin{bmatrix} \hat{R} & \hat{\mathbf{t}} \end{bmatrix} \begin{bmatrix} \mathbf{P}_D \\ 1 \end{bmatrix} \tag{5.1}$$

where $z_C = \mathbf{r_3}^T \mathbf{P}_D + t_z$. In expression (5.1) $\mathbf{r}_3^T$ is the third row of $\hat{R}$, t_z the third component of $\hat{\mathbf{t}}$, $\hat{K}_C$ the estimated intrinsic parameters matrix of C and $(\hat{R}, \hat{\mathbf{t}})$ the roto-translation between C and D estimated by calibration, as described in Chapter 4. The reprojection of the depth data samples $p_D \in Z_C$ on the color image I_C defined on the 2D lattice Λ_C associated to the C-2D reference frame by (5.1) produces a low resolution depth map Z_C^L defined on a subset of Λ_C, since the resolution of I_C is higher than that of Z_D. This is clearly shown by the examples of Figure 5.4.

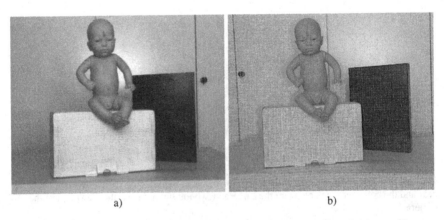

a)   b)

Fig. 5.4 a) Reprojection of the samples acquired by the ToF camera on the color image; b) reprojection of the samples acquired by the Kinect^{TM} range camera on the Kinect^{TM} color image.

The reprojected depth samples need to be interpolated in order to associate a depth value to each pixel of Λ_C and to obtain from Z_C^L a depth map Z_C with spatial resolution equal to that of I_C. Interpolation strategies for this task will be seen in Section 5.2. The artifacts introduced by the interpolation of depth information might however cause errors in the occlusion detection and the projection of some depth samples to wrong locations.

Reprojection (5.1) may also include samples occluded from the camera C point of view. Occluded samples must be removed, since this may be critical for many applications. Removal can be achieved by building a 3D mesh from the depth map Z_D and then by rendering it from the color camera viewpoint. The Z-buffer associated to the rendering can then be used in order to recognize the occluded samples. An example of Z-buffer-based algorithm is presented in [18]. Another possibility [21] is to use the 3D geometry acquired from the range camera D in order to convert I_C to an orthographic image where all the optical rays are parallel. Such a conversion makes trivial the fusion of depth and color data.

A different approach is needed for a representation of depth and color data from the D viewpoint. The color camera provides only 2D information and the acquired pixels can not be directly reprojected to the D viewpoint. In order to obtain a color image from the D viewpoint it is, instead, necessary to build a 3D mesh from the depth map Z_D acquired by D, reproject to I_C the 3D vertices of the mesh visible from the C viewpoint in order to associate a color to the corresponding pixel of Z_D. Therefore from the D viewpoint one may obtain a low resolution color image I_D defined on the 2D lattice Λ_D associated to the D-2D reference frame, and a low resolution depth map Z_D defined on Λ_D. Since in this case the final resolution is the one of the ToF data, there is hardly any resolution change and no special interpolation strategies are needed. Interpolation is only needed for associating a color to the pixels of Z_D occluded from the C viewpoint.

As previously said, representations in term of images and depth maps are typical of 3D video applications, but in other fields such as 3D reconstruction or 3D object recognition, are more common 3D scene descriptions by way of colored point-clouds or textured meshes. In this case, the geometrical description, i.e., the point cloud or the mesh, can be obtained from the depth measurements by either a ToF or a Kinect^{TM} range camera. The color of each acquired point or mesh triangle can then be obtained by projecting on the color camera image the point cloud points or the triangle vertexes by (5.1). After computing the point coordinates on the camera image, the actual color value can be computed from the closest image samples by image interpolation (e.g., bilinear, bicubic, spline, etc.). A similar procedure can be used for the texture coordinates of each 3D mesh triangle.

If the setup features a stereo system S made by two color cameras and a depth camera D, multiple color and depth data streams are available and there are two major possibilities. The first one is to independently apply the previously described methods to each of the two cameras in order to reproject the acquired data on the D viewpoint or on the viewpoint of one of the two cameras. The second possibility is the usage of stereo vision techniques [23] applied to the data acquired from the stereo system S in order to compute the 3D positions of the acquired points. The 3D points obtained from the stereo can then be combined with the 3D data obtained from the depth camera D with respect to a 3D reference system (either the D-3D or the S-3D reference systems) as described in Section 5.4. In this particular case, the adopted approach generally depends on the employed data fusion algorithm.

5.2 Depth data super-resolution: deterministic approaches

As seen in the previous section, calibration and reprojection make available an image I_C (or I_D) and a depth map Z_C^L (or Z_D) referring to the same viewpoint. Depth data typically have lower resolution and poorer edges localization. The better edges localization of color images is therefore suited to improve edges localization and spatial resolution of depth data, an operation often called *depth data super-resolution.*

A first possibility is to extract edge information from the color data by edge detection or segmentation techniques and to use it to assist the interpolation process. A segmentation process divides the color image in a set of regions called *segments* ideally corresponding to the different scene objects. It is reasonable to assume that inside a segment the depth varies smoothly and that sharp depth transitions between different scene objects occur at the boundaries between different segments.

Figure 5.5 shows an example of segments containing the reprojected depth samples Z_C^L from a range camera. The spatial resolution of depth map Z_C^L may be increased to that of Z_C by interpolation schemes computing the missing depth values only from the neighboring depth values inside each segment, i.e., each interpolated depth sample is only function of the samples within the same segment. This is equivalent to confine the low-pass action of the interpolation within each segment and to preserve sharp transitions between different segments.

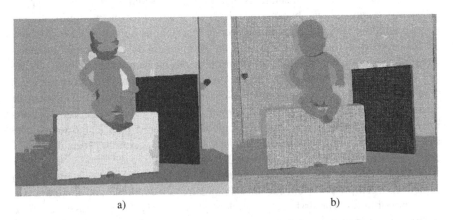

a)   b)

Fig. 5.5 Reprojection of the depth samples on the segmented color image in case of: a) an acquisition system made by a ToF camera and a standard camera; b) an acquisition system made by a Kinect[TM] and a standard camera.

This concept inspires the method of [10], where the depth values are computed by bilinear interpolation of the reprojected depth samples inside each segment. In particular this approach uses some heuristics on the 3D surface shape to compute the position that the depth samples reprojected outside the current segment would have if they lay on an extension of the surface corresponding to the considered segment. The samples predicted in this way are then used to improve interpolation accuracy at edge regions. This approach outperforms standard image interpolation techniques and can produce very accurate depth maps as shown in Figure 5.6. Its performance is limited by two main issues, namely by segmentation errors and by inaccuracies due to depth acquisition or to the calibration between depth and color cameras.

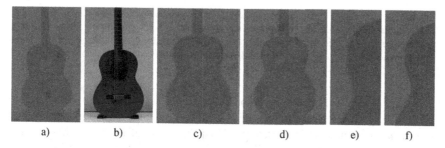

a)   b)   c)   d)   e)   f)

Fig. 5.6 Super-resolution assisted by color segmentation: a) low resolution depth map acquired by a ToF camera; b) image acquired by a color camera; c) high resolution depth map obtained by Lanczos interpolation; d) high resolution depth map obtained with the aid of color information and segmentation by the method of [10]; e) detail of the high resolution depth map of c); f) detail of the high resolution depth map of d).

Segmentation is a very challenging task and, despite the large research activity in this field, currently there are no procedures completely reliable for any scene. As expected, segmentation errors or inaccuracies can lead to wrong depth sample assignments with artifacts in the estimated depth maps of the type shown in Figure 5.7. For example, the reprojected depth sample inside the green circle of Figure 5.7a belongs to the foreground but it has been wrongly assigned to the background because of calibration and segmentation inaccuracies. Figure 5.7b shows how the wrongly assigned depth sample is propagated inside the background by the interpolation algorithm. Inaccuracies in the calibration process or in the acquired depth samples can similarly bring the reprojected depth samples close to the boundaries between segments to "cross" them and to be assigned to wrong regions.

A possible solution to segmentation issues is to replace segmentation by edge detection, which is a simpler and more reliable operation. However, cracks of the edges may allow the reprojected depth samples to be propagated out of the corresponding segments with consequent artifacts in the interpolated depth. False or double edges can affect the interpolation process as well. The artifacts due to the second issue may be reduced by more accurate calibration procedures, as described in Chapter 4. A further possibility is to either exclude or underweight reprojected depth values too close to the edges [10] in order to eliminate unreliable data from the interpolation process. Figure 5.7c shows how this provision eliminates some interpolation artifacts.

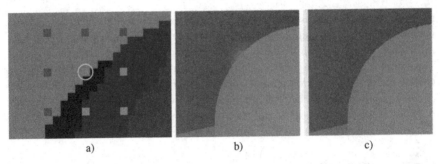

a)   b)   c)

Fig. 5.7 Artifacts due to segmentation and calibration inaccuracies: a) one reprojected depth value (circled in green) has been assigned to the background but it belongs to the foreground; b) high resolution depth map obtained by interpolating the depth data inside the different segments; c) high resolution depth map obtained by method [10] which takes into account also data reliability.

Instead of explicitly detecting color image edges, several recent works [28, 15] resort to range domain filtering techniques guided by color data to interpolate and filter depth information. The bilateral filter [25] is one of the most effective and commonly used edge-preserving filtering schemes. This filter computes the output value $I_f(p^i)$ at pixel p^i with coordinates $\mathbf{p}^i = [u^i, v^i]^T$ as the weighted average of the pixels p^n, $n = 1,...,N$ with coordinates $\mathbf{p}^n = [u^n, v^n]^T$ in a window W^i surrounding p^i as most standard filters used in image processing. The key difference with respect to standard approaches is that the weights do not depend only on the spatial distance between p^i and each p^n in W^i but also on the difference between the color value of p^i and p^n. More precisely $I_f(p^i)$ is computed as the followings:

$$I_f(p^i) = \frac{1}{n_f} \sum_{p^n \in W^i} G_s(p^i, p^n) G_r(I(p^i), I(p^n)) I(p^n) \tag{5.2}$$

where:

$$n_f = \sum_{p^n \in W^i} G_s(p^i, p^n) G_r(I(p^i), I(p^n)) \tag{5.3}$$

Expression (5.2) shows that the filter output at p^i is the weighted average of the image samples in W^i with weights obtained as product of a function G_s depending on the spatial distance between p^i and p^n with a function G_r depending on the color difference between $I(p^i)$ and $I(p^n)$. The spatial weighting function G_s is the standard Gaussian function used in low-pass filtering, i.e.:

$$G_s(p^i, p^n) = \frac{1}{2\pi\sigma_s^2} e^{-\frac{(u^i - u^n)^2 + (v^i - v^n)^2}{2\sigma_s^2}} \tag{5.4}$$

In [25] the color weighting term G_r is also a Gaussian function applied to the difference between the color of the two samples:

$$G_r(p^i, p^n) = \frac{1}{2\pi\sigma_r^2} e^{-\frac{\delta(I(p^i), I(p^n))^2}{2\sigma_r^2}} \tag{5.5}$$

where $\delta(I(p^i), I(p^n))$ is a suitable measure of the difference between the colors of the two samples p^i and p^n. It can be computed, for instance, as the euclidean distance between the colors of p^i and p^n in the CIELab color space [7].

In the considered setup where one has available a low resolution depth map Z_D acquired by a depth camera D and a high resolution color image I_C acquired by a color camera C, a modified version of the bilateral filter allows to obtain a high resolution depth map defined on lattice Λ_C. As noted in Section 5.1, the reprojection of of Z_D on Λ_C gives a low resolution sparse depth map Z_C^L defined on a subset $\Gamma_L \triangleq \{p^l, l = 1, ..., L\}$ of Λ_C. The high resolution color image I_C associates a color value $I_C(p^i)$ to each sample $p^i \in \Lambda_C$. Two different approaches can be considered in order to apply bilateral filtering .

The first one is to compute the filter output $I_f(p^i)$ for each sample of the high resolution depth map using only the samples $I_f(p^l)$ that are inside the window W^i and have an associated depth value, i.e., Equation (5.2) can be rewritten as:

$$Z_{C,f}(p^i) = \frac{1}{\overline{n}_f^i} \sum_{p^l \in \overline{W}^i} G_s(p^i, p^l) G_r(I_C(p^i), I_C(p^l)) Z_C^L(p^l) \tag{5.6}$$

where $\overline{W}^i = W^i \cap \Gamma_L$ is the set of the low resolution depth map points falling inside window W^i, and the normalization factor $\overline{n}_f^i$ is computed as

$$\overline{n}_f^i = \sum_{p^l \in \overline{W}^i} G_s(p^i, p^l) G_r(I_C(p^i), I_C(p^l)) \tag{5.7}$$

It is important to note how in this case the range weighting factor G_r is not computed on the depth values but as the difference between the color of p^i and of p^l on the corresponding color image C acquired by the color camera. This approach is commonly called *cross bilateral filtering* to underline that the range weights come from a domain different from the one of the filtering operation, i.e. the filtering concerns depth data and G_r is computed from associated color data. Figure 5.8 shows an example of the results that can be achieved by this approach, note how the exploitation of the high resolution color information through the range weighting factor G_r allows to correctly locate and preserve the edges.

The second possibility is given by the following two-steps procedure:

a) interpolate the depth map Z_C^L by a standard image interpolation algorithm (e.g. spline interpolation) in order to obtain a high resolution depth map Z_C defined on the high resolution lattice Λ_C;
b) obtain the final high resolution depth map $Z_{C,f}$ by applying the bilateral filter to Z_C with the range weighting factor G_r computed as before, i.e., as the difference between the color of p^i and of p^n on the corresponding color image I_C, i.e.:

$$Z_{C,f}(p^i) = \frac{1}{n_f^i} \sum_{p^n \in W^i} G_s(p^i, p^n) G_r(I_C(p^i), I_C(p^n)) Z_C(p^n) \qquad (5.8)$$

with $n_f^i = \sum_{\mathbf{p}^n \in W^i} G_s(p^i, p^n) G_r(I_C(p^i), I_C(p^n))$.

Note that after the interpolation of step *a*) depth and color are defined on the same lattice, and this allows to use all the samples of window W^i for the interpolation. A similar approach has been employed in [15] with the non-local means filter [6] in place of the bilateral filter. The main difference between non-local means and bilateral filter is that the range weighting factor of the former is computed from two windows surrounding the two samples p^i and p^n instead of just from the two samples p^i and p^n. Furthermore this approach explicitly handles depth data outliers, which are quite common in the data acquired by current depth cameras.

a)   b)

Fig. 5.8 Bilateral upsampling of depth data from a ToF camera (a) and from Kinect^{TM} (b).

5.3 Depth data super-resolution: probabilistic approaches

Probabilistic approaches offer an interesting alternative for depth data super-resolution. Before describing this family of methods, it is worth introducing some basic concepts and notation. A depth map Z can be considered as the realization of a random field $\mathscr{Z}$ defined over a 2D lattice Λ_Z. The random field $\mathscr{Z}$ can be regarded as the juxtaposition of the random variables $Z^i \triangleq \mathscr{Z}(p^i)$, with p^i being a pixel of Λ_Z. The neighbors of p^i are denoted as $p^{i,n} \in N(p^i)$, where $N(p^i)$ is a neighborhood of p^i. Neighborhood $N(p^i)$ can be either the 4-neighborhood, the 8-neighborhood or even a generic window W^i centered at p^i. The random variable Z^i assumes values in the discrete alphabet $z^{i,j}$, $j = 1, ..., J_i$ in a given range $\mathscr{R}^i$. The specific realization of Z^i is denoted as z^i.

In the considered case the high resolution version of the depth distribution can be seen as a Maximum-A-Posteriori (MAP) estimate of the most probable depth distribution $\hat{Z}$ given the acquired depth measurements M^1, i.e.:

$$\hat{Z} = \arg\max_{Z \in \mathscr{Z}} P(Z|M) \tag{5.9}$$

where $\mathscr{Z}$ is the random field of the framed scene depth distribution and Z is a specific realization of $\mathscr{Z}$. The random field $\mathscr{Z}$ is defined on a lattice Λ_Z that can be either the depth camera lattice Λ_D or the color camera lattice Λ_C or an arbitrary lattice different from the source data lattices (e.g., a lattice associated to a *virtual* camera placed between D and C). The choice of this lattice is a first major design decision. In this section we assume to operate on a suitable high resolution lattice (e.g., Λ_C) where depth and color data have been reprojected and interpolated as described in Sections 5.1 and 5.2. The random field $\mathscr{Z}$ is therefore made by a set of random variables $Z^i \triangleq \mathscr{Z}(p^i), p^i \in \Lambda$.

From Bayes rule, Equation (5.9) can be rewritten as

$$\hat{Z} = \arg\max_{Z \in \mathscr{Z}} \frac{P(M|Z)P(Z)}{P(M)} = \arg\max_{Z \in \mathscr{Z}} P(M|Z)P(Z) \tag{5.10}$$

where in the considered case $P(M|Z) = P(Z_D|Z)$ is the likelihood of the measurements made by the depth camera D given scene depth distribution Z, and $P(Z)$ is the prior probability of the scene depth distribution. Field $\mathscr{Z}$ can be modeled as a Markov Random Field (MRF) defined on the considered lattice. Markov Random Fields extend the concept of Markov chains to regular 2D fields such as images or depth maps. A detailed presentation of the theory behind these representations is out of the scope of this book but can be found in many places, e.g. [17]. The MRF are suited to model the relationships between neighboring pixels in the high resolution lattice, i.e., the typical structure of depth information made by smooth regions separated by sharp edges.

Let us recall that by definition $\mathscr{Z}$ is a MRF if

$$\begin{aligned} P(Z^i|Z^n : Z^n &\triangleq \mathscr{Z}(p^n), \forall p^n \in \Lambda_Z \setminus \{p^i\}) = \\ P(Z^i|Z^n : Z^n &\triangleq \mathscr{Z}(p^n), \forall p^n \in N(p^i)), \forall p^i \in \Lambda_Z \end{aligned} \tag{5.11}$$

in which $N(p^i)$ is a suitable neighborhood of p^i. It is possible to demonstrate [17] that $\mathscr{Z}$ is characterized by a Gibbs distribution. Therefore the MAP problem of Equation (5.10) after some manipulation can be expressed as the minimization of energy function

$$U(Z) = U_{data}(Z) + U_{smooth}(Z) = \sum_{p^i \in \Lambda_Z} V_{data}(p^i) + \sum_{p^i \in \Lambda_Z} \sum_{p^n \in N(p^i)} V_{smooth}(p^i, p^n) \tag{5.12}$$

---

[1] In this section the set of depth measurements M comes from a single depth camera, i.e., either a Kinect[TM] or a ToF camera, while in Section 5.4 depth data come from two different depth cameras of which one is a stereo vision system.

Energy $U(Z)$ it is the sum of two energy terms: a *data term* U_{data} modeling the probability that depth assumes a certain value at p^i, and a *smoothness term* U_{smooth} modeling the dependency of the depth value at p^i from its neighbors $p^n \in N(p^i)$. The first term typically accounts for the depth camera measurements, while the second term models the fact that depth maps are made by smooth regions separated by sharp edges. A simple possibility is to just use U_{smooth} to enforce the smoothness of the depth map. In this way the available high resolution color information would not be exploited and an interesting alternate option used in several approaches [9, 20, 14] is taking into account color information in the construction of the MRF associated to the depth data. As shown in Figure 5.9, the basic idea is that the dependency of Z^i from its neighbors $Z^n \triangleq \{ \mathcal{Z}(p^n), p^n \in N(p^i) \}$ is stronger when the samples have similar colors and weaker when the samples have different colors. This models the fact that samples with similar colors are more likely to lie on the same object and thus to have similar depth values, while depth information edges are probably aligned with color data edges.

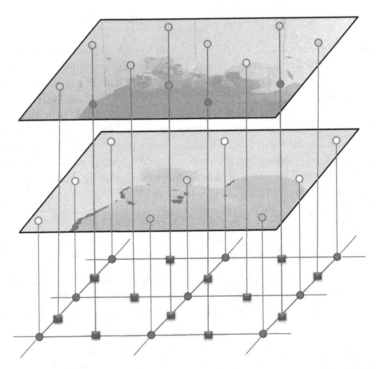

Fig. 5.9 Pictorial illustration of the energy functions used in depth super-resolution: the V_{data} term (shown in blue) depends on the depth camera measurements, while the V_{smooth} term affects the strength of the links (in red) representing the dependency of each depth sample from its neighbors.

Typically the data term U_{data} is the sum of a set of potentials $V_{data}(p^i)$, one for each sample of the lattice (the blue circles in Figure 5.9). The potential $V_{data}(p^i)$

based on the depth camera measurements can be expressed as a function of the difference between the considered value and value $Z_D(p^i)$ actually measured by the depth camera. All the proposed approaches use an expression of the form

$$V_{data}(p^i) = k_1[z^i - Z_D(p^i)]^2 \qquad (5.13)$$

where k_1 is a constant weighting factor used to balance the data term with respect to the smoothness term U_{smooth}.

The smoothness term U_{smooth} modeling the dependency of each sample from its neighbors (represented by the links with the red squares in Figure 5.9) can be expressed as the sum of a set of potentials $V_{smooth}(p^i, p^n)$ defined on each couple of neighboring points on the considered lattice. The potentials $V_{smooth}(p^i, p^n)$ can be defined in different ways, but the most common solution is to model them as a function of the difference between the considered value z^i and its neighbors z^n, i.e.:

$$V_{smooth}(p^i, p^n) = V_{smooth}(z^i, z^n) = w_{i,n}[z^i - z^n]^2 \qquad (5.14)$$

where $w_{i,n}$ are suitable weights modeling the strength of the relationship between close points. They are the key element for this family of approaches: weights $w_{i,n}$ not only model the depth map smoothness but also provide the link between the depth data and the associated color image acquired by the standard camera. If z^i and z^n are associated to samples of similar color on the camera image they probably belong to the same region and thus there should be a strong link between them. If they correspond to samples of different color the smoothness constraint should be weaker. The simplest way to account for this rationale is to compute the weight $w_{i,n}$ linking z^i to z^n as a function of the color difference between the associated color samples $I_C(p^i)$ and $I_C(p^n)$, for instance by an exponential function [9, 14]:

$$w_{i,n} = e^{-k_2 \delta(I_C(p^i), I_C(p^n))^2} \qquad (5.15)$$

where $\delta(I_C(p^i), I_C(p^n))$ is a measure of the color difference between the two samples at location p^i and p^n (e.g., in the simplest case, the absolute value of the difference between their intensity values) and k_2 is a tuning parameter. Needless to say, this simple model can be easily extended in order to feed the probabilistic model with many other clues, such as the segmentation of the color image, the output of an edge detection algorithm, or other image characteristics [20].

Depth distribution $\hat{Z}$ is finally computed by finding the depth values that maximize the Maximum-A-Posteriori (MAP) probability (5.12) modeled by the MRF. This problem can not be subdivided into smaller optimization problems and needs to be solved by complex global optimization algorithms, such as Loopy Belief Propagation [22] or Graph-cuts [4]. The theory behind these optimization methods, rather general and powerful, can be found in [17, 3, 23, 24]. The choice of the specific optimization algorithm is another important element for the depth interpolation algorithm.

5.4 Fusion of data from a depth camera and a stereo vision system

Another interesting possibility to overcome the data limitations of current depth cameras and to provide both color and 3D geometry information is to use an acquisition setup made by a range camera D (either a ToF camera or a Kinect^{TM}) and a couple of color cameras used as an autonomous stereo vision system S. In this case, both the stereo vision system S and the range camera D are able to independently estimate the 3D scene geometry. Moreover each camera of the stereo vision system provides color information.

The quantities that characterize such an acquisition system are:

- the color images pair $I_S = \{I_L, I_R\}$ acquired by the stereo system S (being I_L the image acquired by the left camera L and I_R the image acquired by the right camera R). I_L is defined on lattice Λ_L associated to the L-2D reference system, and I_R is defined on lattice Λ_R associated to the R-2D reference system;
- The depth map Z_S estimated by the stereo vision system from the color images I_S, conventionally defined on lattice Λ_L;
- The full information $F_D = \{A_D, Z_D, (B_D)\}$ acquired by the depth camera D, where A_D is an amplitude image, Z_D a depth map and B_D an intensity image (A_D and B_D may not be made available by all the commercial products). All these quantities are defined on lattice Λ_D associated to the D-2D reference system.

Calibration (that can be performed as described in Chapter 4) allows to spatially relate the 3D geometry information provided by D and the 3D geometry and color information provided by S. Calibration is both a necessary and a critical step for the fusion of the data acquired by D and S.

The rationale behind the fusion of data from a depth camera and a stereo vision system is to provide 3D geometry estimates with characteristics of accuracy, precision, depth resolution, spatial resolution and robustness superior in all or some aspects to those of the original data from the range camera or the stereo system. Most of the proposed methods were originally tailored to the case of ToF cameras, as the introduction of the Kinect^{TM} is very recent and the combination of this device with stereo systems is still a quite unexplored field. Anyway, most of the actual approaches can be generally adapted to the case of a Kinect^{TM}.

The fusion of depth information coming from a range camera D and a stereo vision system S, seen in the previous section, can be formalized by the MAP probabilistic approach (5.9) that can be rewritten by Bayes rule as in (5.10).

In this case the data M measured by the acquisition system are the full information F_D acquired by D and the color images I_S acquired by S. $\mathcal{Z}$ is the random field of the framed scene depth distribution, defined on Λ_Z, which as in the case of super-resolution can be Λ_D, Λ_S or another lattice (e.g., a lattice associated to a virtual camera placed between L and D). The choice of lattice Λ_Z is a major design decision when fusing data from S and D. Approaches like [30, 29, 31] adopt $\Lambda_Z = \Lambda_S$, while other approaches such as [8] adopt $\Lambda_Z = \Lambda_D$. The choice of Λ_S leads to high resolution depth map estimates (i.e., to depth maps with the resolution of stereo images I_S), while the choice of Λ_D leads to low resolution depth map estimates. It is worth noting that although the choice of Λ_D does not improve the depth map resolution, it may also be of interest since it leads to more accurate estimates, as explained later in this section.

Another fundamental characteristic that defines the random field $\mathcal{Z}$ is the range of values that each random variables $Z^i = \mathcal{Z}(p^i)$ can assume. Each random variable Z^i assumes values in the discrete alphabet $z^{i,j}$, $i = 1, 2, .., J^i$ in a given range $R^i \subset [z_{min}, z_{max}]$ where z_{min} is the nearest measurable distance (e.g., $500[mm]$ for a Kinect^{TM}) and z_{max} is the furthest measurable distance (e.g., $5 \times 10^3[mm]$ for a ToF camera with modulation frequency $f_{mod} = 30[MHz]$). Note that also the distances measured by a stereo vision system are bounded by the choice of a minimum and a maximum disparity value. The choice of z_{min} and z_{max} is the second major design decision which has to account for the characteristics of D, S and of the framed scene (prior knowledge of minimum and maximum depth of the scene to be acquired is rather common). It is also worth pointing that alphabet values $z^{i,j}$ and alphabet cardinality J^i depend on p^i.

The likelihood term $P(I_S, F_D|Z)$ accounts for the depth information acquired by both D and S. Under the common assumption that the measurement likelihoods of the two 3D acquisition systems (in this case D and S) are independent [13, 8, 30, 29, 31], the likelihood term can be rewritten as

$$P(I_S, F_D|Z) = P(I_S|Z)P(F_D|Z) \qquad (5.16)$$

where $P(I_S|Z)$ is the likelihood of the S measurements and $P(F_D|Z)$ the likelihood of the D measurements. Therefore in this case the MAP problem of Equation (5.10) can be rewritten as:

$$\hat{Z} = \arg\max_{Z \in \mathcal{Z}} P(Z|I_S, F_D) = \arg\max_{Z \in \mathcal{Z}} P(I_S|Z)P(F_D|Z)P(Z) \qquad (5.17)$$

Each one of the two likelihoods $P(I_S|Z)$ and $P(F_D|Z)$ can be computed independently for each pixel $p^i \in \Lambda_Z$. Let us consider first $P(I_S|Z)$ and observe that from (1.15) the left camera pixel p^i with coordinates $\mathbf{p}^i = [u^i, v^i]^T$ and the depth values $z^{i,j}$ identify a set of 3D points $P^{i,j}$ with coordinates $\mathbf{P}^{i,j} = [x^{i,j}, y^{i,j}, z^{i,j}]^T$ $j = 1, ..., J_i$. Such points are projected on the conjugates pairs $(p_L^{i,j}, p_R^{i,j})$. Therefore the likelihood $P(I_S|Z)$ can be regarded as the set of $P(I_S|P^{i,j}) \triangleq P(I_S/Z^i = z^{i,j})$ $j = 1, 2, ..., J_i$. The stereo measurements likelihood for $P^{i,j}$ can be computed by comparing the I_L image in $W_L^{i,j}$, a window centered at $p_L^{i,j}$, and the I_R image in $W_R^{i,j}$, a window centered at $p_R^{i,j}$. In principle the two windows are very similar if $z^{i,j}$ is close to the correct depth value, while they are different if $z^{i,j}$ is different from the correct depth value. The likelihood therefore assumes high values if the windows $W_L^{i,j}$ and $W_R^{i,j}$ have a high similarity score and low values if the windows have low similarity score. Most stereo vision systems compute a cost function $c_{i,j} = \mathscr{C}(p_L^{i,j}, p_R^{i,j})$ representing the similarity between the two windows and the likelihood of a set of points $P^{i,j}$,

$i = 1, 2, ..., J_i$ can be computed as a function of $c_{i,j}$, for instance by an exponential model as:

$$P(I_S | P^{i,j}) \propto e^{-\frac{c_{i,j}}{\sigma_S^2}} \qquad (5.18)$$

where σ_S is a normalization parameter, experimentally found on the basis of the color data variance. The cost function $\mathscr{C}$ can be any of the different functions used in stereo vision techniques [23], such as a simple Sum of Squared Differences (SSD):

$$c_{i,j} = \sum_{p_L^n \in W_L^{i,j}, p_R^n \in W_R^{i,j}} [I_L(p_L^n) - I_R(p_R^n)]^2 \qquad (5.19)$$

The windows $W_L^{i,j}$ and $W_R^{i,j}$ can be for instance rectangular windows centered at $p_L^{i,j}$ and $p_R^{i,j}$ respectively. Stereo cost computation is pictorially shown in Figure 5.10, where the cost value $c_{i,j}$ relative to a specific realization z^i of Z^i is simply denoted as c_i.

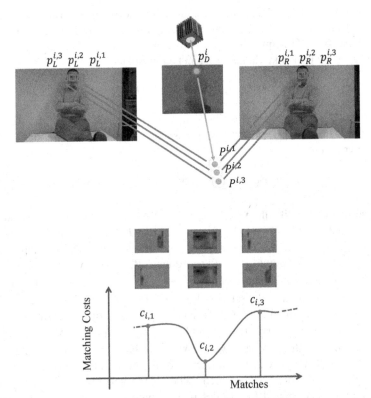

Fig. 5.10 Pictorial illustration of stereo costs computation. Different 3D points associated to different depth measurements (green dots) project to different pairs of possible conjugate points. For each pair of possible conjugate points a matching cost (shown in the plot) is computed by locally comparing the stereo images in windows (red rectangles) centered at the relative conjugate points.

The depth cameras likelihood $P(F_D|Z)$ for a given point $P^{i,j}$ can be regarded as the set of $P(F_D|P^{i,j}) \triangleq P(F_D|Z^i = z^{i,j}), j = 1,2,..,J^i$ and can be computed by analyzing the measurement errors of depth cameras. Let us recall that depth cameras measurements are generally characterized by random errors distributed along the cameras optical rays (Chapters 2 and 3). For example, in the case of a ToF camera, the depth random error is characterized by a Gaussian distribution along the optical rays (at least for the measured points far from depth discontinuities) with standard deviation σ_P that can be obtained from (2.11), as hinted by Figure 5.11. For the Kinect^{TM} case, the error distribution is less regular, but the error is always

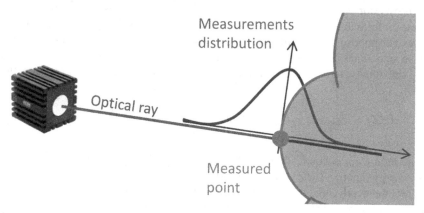

Fig. 5.11 Pictorial illustration of distance measurement error (blue Gaussian distribution) along the depth camera optical ray (red line).

distributed along the Kinect^{TM} IR camera optical rays.

Since in both the cases of ToF cameras or Kinect^{TM} the error is distributed along the optical rays, the D likelihood terms $P(F_D|Z^i = z^{i,j})$ can be independently considered for each point of lattice $\Lambda_Z = \Lambda_D$. The choice of Λ_D for the estimated depth random field allows to exploit this property (and this is why the choice of Λ_D as lattice for $\mathscr{Z}$ is not naive).

Let us suppose that the depth estimation error is distributed as a Gaussian. The depth camera measurements likelihood for a set of 3D points $P^{i,j}$, $j = 1,2,...J_i$ relative to $p^i \in \Lambda_Z$ can be computed as:

$$P(F_D|P^{i,j}) \propto e^{-\frac{z^{i,j} - Z_D(p_D^i)}{\sigma_D^2}} \tag{5.20}$$

where $z^{i,j}$ is the depth coordinate of $P^{i,j}$, $Z_D(p_D^i)$ is the measured depth coordinate, p_D^i is the pixel of lattice Λ_D associated to $P^{i,j}$. Parameter σ_D is a suitable function of the depth data variance and noise, e.g., in [8] it depends both on the variance of the 3D point positions σ_P and on the local variance of the depth measurements inside a window centered at p_D^i.

The last term of Equation (5.17) to model is prior $P(Z)$ which can be modeled in two ways, namely by an independent random variable approach or by a MRF approach, as seen next.

5.4.1 Independent random variable approach

The first approach (explicitly adopted in [8] and implicitly in [27]) models $\mathscr{Z}$ as a juxtaposition of independent random variables $Z^i \triangleq \mathscr{Z}(p^i)$, $p^i \in \Lambda$, where Λ is the considered lattice (either Λ_S, Λ_D or another lattice). These variables are characterized by prior probability $P(z^{i,j}) \triangleq P(Z^i = z^{i,j})$, which for instance in [8] is a discrete uniform distribution in $[z_{min}, z_{max}]$. The uniform probability distribution model does not impose any specific structure to the scene depth distribution (e.g., piecewise smoothness). In this case the MAP problem of (5.17) can be simplified and the smoothness term can be removed from the minimization of the energy function $U(Z)$, i.e.:

$$U(Z) = U_{data}(Z) = U_S(Z) + U_D(Z) = \sum_{p^i \in \Lambda_Z} V_S(p^i) + \sum_{p^i \in \Lambda_Z} V_D(p^i) \qquad (5.21)$$

Data term U_{data} accounts both for the contribution of the stereo system S (through the energy function $U_S(Z)$ and the corresponding potentials $V_S(p^i)$) and the contribution of the depth camera (through $U_D(Z)$ and the corresponding potentials $V_D(p^i)$). More precisely, $V_S(p^i)$ depends on the cost function $\mathscr{C}$ of the stereo system. For instance, if the model of Equation (5.18) is used for the cost function, $V_S(p^i)$ can be computed as:

$$V_S(p^i) = \frac{1}{\sigma_S^2} c_i \qquad (5.22)$$

The potentials $V_D(p^i)$ instead depend on the measurements of the depth camera, e.g., using the model of Equation (5.20), $V_D(p^i)$ is:

$$V_D(p^i) = \frac{[z^i - Z_D(p_D^i)]^2}{\sigma_D^2} \qquad (5.23)$$

Therefore from (5.22) and (5.23) expression (5.21) can be rewritten as

$$U(Z) = \sum_{p^i \in \Lambda_Z} \left[\frac{c_i}{\sigma_S^2} + \frac{[z^i - Z_D(p_D^i)]^2}{\sigma_D^2} \right] \qquad (5.24)$$

Equation (5.24) can be optimized independently for each random variable of the estimated random field by a Winner-Takes-All (WTA) approach. The WTA approach, for each random variable in Λ_Z, picks the depth value z^i which maximizes the correspondent energy term in (5.24). Approaches based on independence assumptions consider each random variable of the random field independently and this leads to a lack of global knowledge about the estimated random field. Such approaches, often called *local methods*, generally aim at improving the accuracy and the precision of the estimated depth distribution with respect to both the original estimates of D and S, without applying any global regularization model. Figure 5.12 reports some results obtained from the local method of [8].

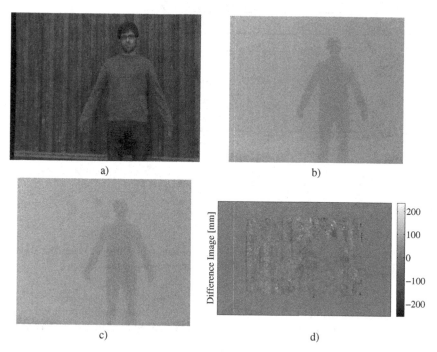

a)   b)

c)   d)

Fig. 5.12 Results of the application of the local fusion approach of [8]: a) acquired left color image; b) depth map acquired by the ToF camera; c) estimated depth map after the fusion; d) difference between the depth map acquired by the ToF camera and the depth map estimated by the fusion approach.

5.4.2 MRF modeling approaches

Similarly to the case of Section 5.3, also in this case the prior probability $P(Z)$ can be modeled as a Markov-Random-Field (MRF). As previously explained, this model imposes piecewise smoothness to the estimated depth map. In this case $\mathscr{Z}$ is characterized by a Gibbs distribution, and prior $P(Z)$ can be computed as

$$P(Z) \propto \prod_{p^i \in \Lambda_Z} \prod_{p^n \in N(p^i)} e^{-V_{smooth}(z^i, z^n)} \qquad (5.25)$$

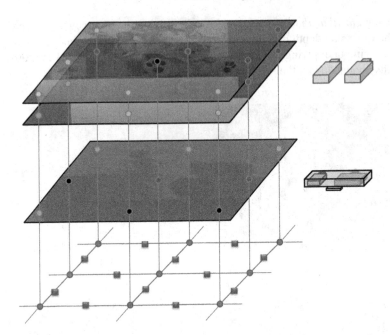

Fig. 5.13 Markov Random Field used for the fusion of data from a depth camera and a stereo system. The data term depends on both depth camera and stereo vision measures.

where z^i is a realization of $Z^i = \mathcal{Z}(p^i)$ and $N(p^i)$ is a suitable neighborhood of p^i. Also in this case the MAP problem of (5.17) can be solved by minimizing an energy function $U(Z)$ of the following form:

$$
\begin{aligned}
U(Z) &= U_{data}(Z) + U_{smooth}(Z) \\
&= U_S(Z) + U_D(Z) + U_{smooth}(Z) \\
&= \sum_{p^i \in \Lambda_Z} V_S(p^i) + \sum_{p^i \in \Lambda_Z} V_D(p^i) + \sum_{p^i \in \Lambda_Z} \sum_{p^n \in N(p^i)} V_{smooth}(p^i, p^n)
\end{aligned}
\tag{5.26}
$$

in which $V_S(Z)$ and $V_D(Z)$ can be computed according to (5.22) and (5.23) respectively and $V_{smooth}(p^i, p^n)$ forces the piecewise smoothness of depth data.
A typical expression for $V_{smooth}(p^i, p^n)$ is:

$$
V_{smooth}(p^i, p^n) = \min \left\{ [z^i - z^n]^2, T_h^2 \right\}
\tag{5.27}
$$

in which T_h is a robustness threshold that avoids $V_{smooth}(p^i, p^n)$ to take too large values near discontinuities. Note how expression (5.26) differs from (5.12) since in Equation (5.12) the prior term depends only on depth information and the color information affects the smoothness term, while in Equation (5.26) both depth and color enter the data term, as the comparison of Figures 5.9 and 5.13 pictorially hints. Equation (5.26) can be rewritten as

$$U(Z) = \sum_{p^i \in \Lambda_Z} \left[\frac{c_i}{\sigma_S^2} + \frac{[z^i - Z_D(p_D^i)]^2}{\sigma_D^2} + \sum_{p^n \in N(p^i)} \min\left\{ [z^i - z^n]^2, T_h^2 \right\} \right] \quad (5.28)$$

With respect to (5.24), this energy function takes into account both the fidelity with respect to the D and S measurements and the smoothness of the estimated depth distribution. The maximization of (5.28) cannot be done independently for each random variable of the random field, but it requires the same techniques used for the maximization of (5.12), e.g., Loopy Belief Propagation or Graph-Cuts.

Examples of techniques based on MRF in order to fuse data from a ToF camera and from a stereo vision system are presented in [30, 29, 31]. Some results from [31] are shown in Figure 5.14. This kind of approaches allows to improve the accuracy

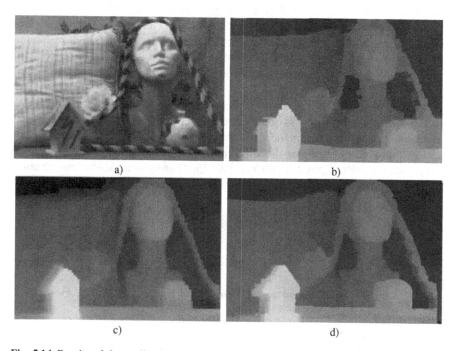

Fig. 5.14 Results of the application of the global fusion approach of [31] (courtesy of the authors): a) acquired left color image; b) depth map acquired by the ToF camera and interpolated; c) depth map produced by the stereo vision algorithm; d) depth map after application of the fusion algorithm.

and the precision of the range camera and of the stereo vision measurements thanks to the regularization due to the smoothness term and to obtain a final depth estimate at the high spatial resolution of the stereo vision data.

5.5 Conclusion and further reading

The fusion of data from sensors of different nature is a classical problem well for-malized in [12]. This chapter addresses the specific case of the fusion of data from a range camera and one or two color cameras. The fundamental instruments for treat-ing these two problems are the bilateral filter [25] and the Markov-Random-Field (MRF) framework [17, 3]. Other techniques, such as non-local means filter [15, 6] and Conditional-Random-Fields (CRF) [26] have also been adopted in this context. Examples of the main methods for depth super-resolution through the fusion of data from a range camera and a standard color camera are in [28, 10, 9, 15, 20, 14], and examples of the main approaches for the fusion of data from a range camera and a stereo vision system are in [30, 29, 31, 8, 27, 16, 11, 28].

In general it can be said that data fusion by either local or global approaches im-proves the accuracy and the precision of both depth camera and stereo measure-ments. The spatial resolution of the fused data may be either the high resolution of the stereo data or the low resolution of data from the depth camera. Global ap-proaches allow to obtain better results at the expenses of a computational load in-crease.

A deep understanding of the methods proposed for the fusion of range camera and stereo data requires awareness of the current stereo vision methods. A fundamental component of the MRF approaches is the optimization of the global energy typ-ical of this kind of problems. Such minimization can be done by Loopy-Belief-Propagation (LBP) [2], by Graph-Cuts (GC) [5, 4] or by other algorithms. A com-prehensive survey of optimization algorithms suitable to this kind of problems can be found in [24].

References

1. T. D. Arun Prasad, K. Hartmann, W. Weihs, S.E. Ghobadi, and A. Sluiter. First steps in enhancing 3d vision technique using 2d/3d sensors, 2006.
2. C. M. Bishop. *Pattern Recognition and Machine Learning (Information Science and Statis-tics).* Springer, 2007.
3. A. Blake, P. Kohli, and C. Rother, editors. *Markov Random Fields for Vision and Image Processing.* MIT Press, 2011.
4. Y. Boykov and V. Kolmogorov. An experimental comparison of min-cut/max-flow algorithms for energy minimization in vision. *IEEE Transactions on Pattern Analysis and Machine Intel-ligence,* 26:359–374, 2001.
5. Y. Boykov, O. Veksler, and R. Zabih. Fast approximate energy minimization via graph cuts. *IEEE Transactions on Pattern Analysis and Machine Intelligence,* 23:1222–1239, 2001.
6. A. Buades, B. Coll, and J.-M. Morel. A non-local algorithm for image denoising. In *Proc. IEEE Computer Society Conf. Computer Vision and Pattern Recognition CVPR 2005,* vol-ume 2, pages 60–65, 2005.
7. Recommendations on uniform color spaces, color difference equations, psychometric color terms. Supplement No.2 to CIE publication No. 15 (E.-1.3.1) 1971/(TC-1.3.), 1978.
8. C. Dal Mutto, P. Zanuttigh, and G. M. Cortelazzo. A probabilistic approach to tof and stereo data fusion. In *Proceedings of 3DPVT,* Paris, France, May 2010.

9. J. Diebel and S. Thrun. An application of markov random fields to range sensing. In *Proceedings of Conference on Neural Information Processing Systems (NIPS)*, Cambridge, MA, 2005. MIT Press.

10. V. Garro, C. Dal Mutto, P. Zanuttigh, and G. M. Cortelazzo. A novel interpolation scheme for range data with side information. In *CVMP*, pages 52–60, nov. 2009.

11. S. A. Gudmundsson, H. Aanaes, and R. Larsen. Fusion of stereo vision and time of flight imaging for improved 3d estimation. *Int. J. Intell. Syst. Technol. Appl.*, 5:425–433, 2008.

12. D.L. Hall and J. Llinas. An introduction to multisensor data fusion. *Proceedings of the IEEE*, 85(1):6–23, jan 1997.

13. C. E. Hernandez, G. Vogiatzis, and R. Cipolla. Probabilistic visibility for multi-view stereo. In *Proc. of CVPR Conf.*, 2007.

14. B. Huhle, S. Fleck, and A. Schilling. Integrating 3d time-of-flight camera data and high resolution images for 3dtv applications. In *3DTV Conference, 2007*, pages 1–4, may 2007.

15. B. Huhle, T. Schairer, P. Jenke, and W. Strasser. Fusion of range and color images for denoising and resolution enhancement with a non-local filter. *Computer Vision and Image Understanding*, 114(12):1336–1345, 2010.

16. K. D. Kuhnert and M. Stommel. Fusion of stereo-camera and pmd-camera data for real-time suited precise, 2006.

17. S. Z. Li. *Markov Random Field Modeling in Image Analysis*. Springer, New York, 3rd edition edition, 2009.

18. M. Lindner, A. Kolb, and K. Hartmann. Data-fusion of pmd-based distance-information and high-resolution rgb-images. In *Signals, Circuits and Systems, 2007. ISSCS 2007. International Symposium on*, volume 1, pages 1–4, july 2007.

19. M. Lindner, M. Lambers, and A. Kolb. Sub-pixel data fusion and edge-enhanced distance refinement for 2d/3d images. *International Journal of Intelligent Systems Technologies and Applications*, pages 344–354, 2008.

20. J. Park, H. Kim, Y.W. Tai, M.S. Brown, and I. Kweon. High quality depth map upsampling for 3d-tof cameras. In *Proceedings of the 13th International Conference on Computer Vision (ICCV2011)*, November 2011.

21. R. Reulke. Combination of distance data with high resolution images. In *Proceedings of Image Engineering and Vision Metrology (IEVM)*, 2006.

22. J. Sun, N. Zheng, and H. Shum. Stereo matching using belief propagation. *IEEE Trans. Pattern Anal. Mach. Intell.*, 25:787–800, 2003.

23. R. Szeliski. *Computer Vision: Algorithms and Applications*. Springer, New York, 2010.

24. R. Szeliski, R. Zabih, D. Scharstein, O. Veksler, V. Kolmogorov, A. Agarwala, M. Tappen, and C. Rother. A comparative study of energy minimization methods for markov random fields with smoothness-based priors. *Pattern Analysis and Machine Intelligence, IEEE Transactions on*, 30:1068–1080, 2008.

25. C. Tomasi and R. Manduchi. Bilateral filtering for gray and color images. In *Computer Vision, 1998. Sixth International Conference on*, pages 839–846, jan 1998.

26. H. M. Wallach. Conditional random fields: An introduction. Technical report, University of Pennsylvania, 2004.

27. Q. Yang, K.H. Tan, B. Culbertson, and J. Apostolopoulos. Fusion of active and passive sensors for fast 3d capture. In *Multimedia Signal Processing (MMSP), IEEE International Workshop on*, 2010.

28. Q. Yang, R. Yang, J. Davis, and D. Nistr. Spatial-depth super resolution for range images. In *Computer Vision and Pattern Recognition, 2007. CVPR '07. IEEE Conference on*, 2007.

29. J. Zhu, L. Wang, J. Gao, and R. Yang. Spatial-temporal fusion for high accuracy depth maps using dynamic mrfs. *IEEE Transactions on Pattern Analysis and Machine Intelligence*, 32:899–909, 2010.

30. J. Zhu, L. Wang, R. Yang, and J. Davis. Fusion of time-of-flight depth and stereo for high accuracy depth maps. In *CVPR*, 2008.

31. J. Zhu, L. Wang, R. Yang, J. E. Davis, and Z. Pan. Reliability fusion of time-of-flight depth and stereo geometry for high quality depth maps. *IEEE Trans. Pattern Anal. Mach. Intell.*, 33:1400–1414, 2011.

Chapter 6
Scene Segmentation and Video Matting Assisted by Depth Data

Detecting the regions of the various scene elements is a well-known computer vision and image processing problem called *segmentation*. Scene segmentation has traditionally been approached by way of (single) images, since they are the most common way of representing scenes. Despite the huge amount of research devoted to this task and the great number of adopted approaches [23], some of which based on powerful techniques such as graph-cuts [12] and mean-shift clustering [8, 21], image segmentation, i.e., scene segmentation by means of color information alone, remains a very challenging task. This is due to the fact that image segmentation is an ill-posed problem essentially because color data do not always contain enough information to disambiguate all the scene objects. For example, the segmentation of an object in front of a background of very similar color remains a very difficult task even for the best segmentation algorithms. Also objects with very complex texture patterns are difficult to segment, specially if the statistics of their texture is similar to that of the background.

Depth information is a very useful clue for scene segmentation. Segmentation based on depth data usually provides better results than image segmentation because depth information allows to easily divide near and far objects on the basis of their distance from the depth camera. Furthermore, it does not have issues related to object and background color and it can easily handle objects with complex texture patterns. Unfortunately, some scene configurations are critical also for depth information, for example the case of two objects touching each other. Until now, segmentation by means of depth information has received limited attention due to the fact that depth data acquisition was a difficult task, but the introduction of ToF sensors and of the Kinect^{TM} are making this task much easier.

Standard image segmentation methods can be easily applied to the depth maps produced by ToF cameras or by the Kinect^{TM}, but some instrinsic data limitations must be taken into account. First of all, the current resolution of these devices limits the precision of the segmented objects contours. Furthermore, there are scene situations more suited to be handled by depth information and others more suited to color information. For all these reasons the joint usage of color and depth data seems the best scene segmentation option. This chapter assumes color and depth data relative to the same viewpoint, obtained by acquisition setups with calibration and reprojection procedures of the kind introduced in Chapter 4 in order to register color and depth data.

Video matting, i.e. the task of dividing foreground objects from the scene background, is a problem strictly related to scene segmentation with practical relevance in many applications, e.g., film-making, special effects, 3D video, etc.. Considerable research is devoted to this particular task. This chapter in Section 6.1 addresses first the use of ToF cameras and Kinect^{TM} in video matting and then in Section 6.2 in the more general problem of scene segmentation.

6.1 Video matting by color and depth data

Video matting has always been an important task in film-making and video production, but in spite of a considerable research activity and of the existence of various commercial products it remains a challenging task. Accurate foreground extraction is feasible by using a cooperative background, but if the background is not controllable and, for instance, it includes colors similar to the ones of the foreground or moving objects, video matting becomes rather difficult. The video matting problem can be formalized [18] by representing each pixel p^i in the observed color image I_C as the linear combination of a background color $B(p^i)$ and a foreground color $F(p^i)$ weighted by the opacity value $\alpha(p^i)$, i.e.:

$$I_C(p^i) = \alpha(p^i)F(p^i) + [1 - \alpha(p^i)]B(p^i) \qquad (6.1)$$

where $\alpha(p^i) = 1$ for foreground pixels and $\alpha(p^i) = 0$ for background pixels. If we constrain α to assume only 0 or 1 values, the problem is reduced to a binary segmentation task. Many methods also allow fractional α values in order to handle transparent objects and pixels close to the edges that include in their area both foreground and background elements.

The set made by the $\alpha(p^i)$ values of the whole image is called *alpha matte* and its estimate is called *matting problem*. It can be easily seen that it is an underconstrained problem. Assuming the use of a standard three-dimensional color space, for each pixel there are seven unknowns (the α value, and three components of each foreground and background color vector) and only three known values (i.e., the color of the observed image). Standard matting approaches adopt assumptions on background and foreground image statistics and on the user input in order to further constrain the problem. Quite clearly depth cameras may bring further distance data which may be rather valuable for its solution. Foreground objects, in fact, have the basic property of being closer to the camera than the background ones and can be then easily detected by depth information. This explains why video matting is one of the key applications of ToF cameras.

The analysis of the objects depth allows to easily separate background and foreground (even by a simple thresholding of the depth values) within the already underlined limitations of depth cameras. In particular, it is very hard to precisely locate the edges between the two regions from depth data alone, because of the limited resolution of ToF cameras, the edge artifacts introduced by many ToF cameras (and by the Kinect^{TM}). If the depth cameras are assisted by standard cameras a further constraint derives from the limited accuracy of the calibration between the depth camera and the high quality color camera used for video acquisition (usually the target is the matting of the video stream of the color camera). Moreover, as already shown in Chapter 5, the depth camera and the color camera have slightly different optical centers, therefore some background regions visible from the color camera may be occluded from the depth camera point of view and a depth-to-color value association may not be feasible.

A common approach in order to deal with the above issues in the matting context is building a *trimap* from depth information, i.e., dividing the image into a foreground, a background and an uncertainty region. The uncertainty region usually corresponds to the areas near the edges between foreground and background (as shown in Figure 6.1). Standard matting algorithms typically build a trimap under some kind of human supervision, but the availability of depth information allows to solve this task automatically, which is a major advantage for practical applications. The simplest approach for obtaining a trimap is to first threshold the depth map and then erode the foreground and background regions in order to remove the pixels near the boundary. The uncertainty region is usually given by the eroded samples and by the samples without depth value either because of occlusions or because it was not provided by the depth camera. The trimap computation can also include further clues, if available; for instance, the ToF camera confidence map can be used to include also pixels with low confidence depth values in the uncertainty region.

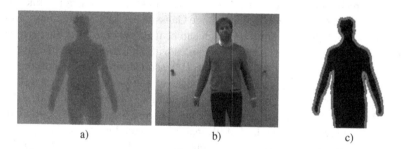

a)   b)   c)

Fig. 6.1 Example of matting trimap: a) depth map acquired by the ToF camera; b) color image acquired by the video camera; c) trimap computed from depth information with black, white and gray pixels referring to foreground, background and uncertainty region respectively.

Another possibility for trimap computation [9] is to assign each sample a probability of being foreground which does not depend from its own depth only but also from the depth of the pixels around it. In this approach each pixel p^i is first assigned a probability of being foreground $P_{fg}(p^i)$ on the basis of the corresponding depth value. The pixels without an associated depth values usually receive a low P_{fg} score in order to model the fact that occluded pixels usually belong to the background (although the lack of a depth measurement may also derive from other causes). Then the likelihood of each pixel being foreground or background is computed as a weighted average of the probabilities of the pixels around it, according to

$$P_{fg}(p^i|W) = \frac{1}{\sum_k w(p^k)} \sum_{p^n \in W^i} w(p^n)P_{fg}(p^n) \qquad (6.2)$$

where W^i is a rectangular window around p^i and $w(p^n)$ are weights usually giving more relevance to the pixels closer to p^i. A Gaussian function is a common choice for weights $w(p^n)$. Note how (6.2) assigns likelihood values close to 0 or 1 to the pixels of the background or foreground regions respectively, and intermediate likelihood values to the pixels near edges. Two thresholds T_l and T_h can be used to assign the samples to foreground, background or uncertainty region, i.e.:

$$p^i \in \begin{cases} \text{foreground} & \text{if } P_{fg}(p^i|W) > T_h \\ \text{background} & \text{if } P_{fg}(p^i|W) < T_l \\ \text{uncertainty region} & \text{if } T_l \leq P_{fg}(p^i|W) \leq T_h \end{cases} \qquad (6.3)$$

As expected, the critical issue for this family of approaches is how to subsequently assign the pixels of the uncertainty region to the background or foreground. As already seen for depth super-resolution in Chapter 5, a possible solution is to apply cross bilateral filtering to the alpha matte, as for the super resolution case [9] seen in Section 5.2, i.e.,

$$\alpha_f(p^i) = \frac{1}{n_f} \sum_{p^n \in W^i} G_s(p^i, p^n)G_r(I(p^i), I(p^n))\alpha(p^n) \qquad (6.4)$$

where W^i is a rectangular window around p^i, n_f is a normalization factor defined as in (5.3), and G_s and G_r are the spatial and range Gaussian weighting functions introduced in Section 5.2. Note that this approach can be used either to filter the existing alpha matte or to assign an α value to the samples without a depth value because of occlusions or missing data in the ToF or Kinect[TM] acquisitions. For the purpose of handling missing data, W^i can be defined as the set of window samples with a *valid* α value (i.e., if the trimap is computed by thresholding depth information, the valid pixels are the ones with an associated depth value). Figure 6.2 shows an example of alpha matte computed by the method of [9] that is based on this approach.

Another possibility is to associate to each image pixel a 4D vector $I_{C,Z}(p^i) = [R(p^i), G(p^i), B(p^i), Z(p^i)]$ with depth as fourth channel to be added to the three standard RGB color channels, and then apply standard matting techniques originally developed for color data to $I_{C,Z}(p^i)$. For example, Wang et al. [27] extend the Bayesian matting scheme originally proposed in [6] for color images by introducing the depth component in the probability maximization scheme. Another interesting idea, also proposed in [27], is weighting the confidence of the depth channel on the basis of the estimated α value in order to give more relevance to the depth of the pixels with α close to 0 or 1 and less relevance to the depth of the pixels in the uncertainty region. This provision takes into account the fact that depth data are less

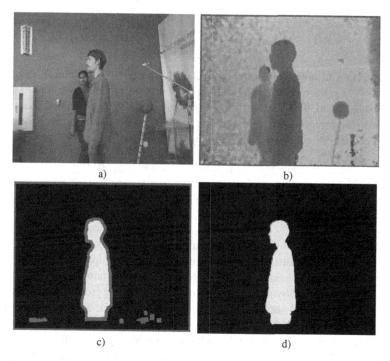

Fig. 6.2 Example of matting by the approach of [9] based on bilateral filtering on color and depth data: a) color image; b) depth map; c) trimap computed on the basis of depth information; d) alpha matte computed by joint bilateral filtering on depth and color data (*courtesy of the authors*).

reliable on the boundary regions between foreground and background due to low resolution and edge artifacts typical of ToF cameras and of the Kinect™.

Poisson matting [22] can be similarly extended in order to include also depth information. In this approach, the color image is first converted to a single channel representation. As in the previous case, it is possible to represent depth as a fourth channel and to include it in the computation of the single channel representation. However depth information is characterized by sharp edges which have a very strong impact on the gradients used by Poisson matting and in this case the obtained results are very similar to the ones obtained from depth information alone. In [27] a confidence map is first built from the depth map. The confidence map is then used in order to derive a second alpha matte which is combined with the one obtained by Poisson matting. A multichannel extension of Poisson matting is proposed in [25] in order to jointly consider the three color channels and the depth channel.

A different class of approaches for the combined segmentation of depth and color data extends the graph-cut segmentation framework to this particular task. These approaches represent the image as a graph $G = \{V, E\}$ where the graph nodes correspond to the pixels and the edges represent the relationships between neighbor pixels (see Figure 6.3). The matting problem can be expressed as the identification of the labeling that minimizes an energy functional of the form:

$$U(\alpha) = \sum_{p^i \in I} V_{data}(\alpha(p^i)) + \sum_{p^i \in I} \sum_{p^n \in N(p^i)} V_s(\alpha(p^i), \alpha(p^n)) \qquad (6.5)$$

It is important to note that these approaches do not allow fractional α values and $\alpha(p^i) \in \{0,1\}$ are binary labels assigning each pixel either to the foreground or to the background. The data term V_{data} of (6.5) models the likelihood that a pixel belongs to the foreground or to the background. It is typically the sum of two terms, one depending on color and one depending on depth, i.e.,

$$V_{data}(\alpha(p^i)) = V_{color}(\alpha(p^i)) + \lambda V_{d,depth}(\alpha(p^i)) \qquad (6.6)$$

where the color term V_{color} can be simply the distance between the pixel color and the mean foreground or background color as in [1] and the depth term V_{depth} can be modeled in a similar way. However V_{depth} must take into account that foreground pixels usually lie at a similar distance from the camera, instead the background pixels can be at different distances from it, as common in complex scenes. In [1] this issue is handled by considering the depth term for foreground pixels only. Better results can be obtained by more complex models of the foreground and background likelihoods. Figure 6.4 shows some results obtained by the approach of [26] that models the two likelihoods as Gaussian Mixture Models (GMM).

Another key issue is that color and depth lie in completely different spaces and it is necessary to adjust their mutual relevance. The proper setting of the weighting constant λ is a challenging task that can be solved through adaptive weighting strategies. In [26] the weights are updated on the basis of the foreground and background color histograms along the spatial and temporal dimensions.

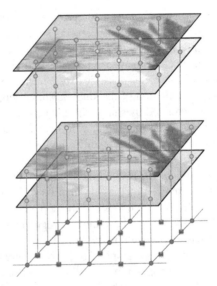

Fig. 6.3 Structure of the Markov Random field used in joint depth and color matting.

The smoothness term V_{smooth} of (6.5) can be built in different ways. In standard graph-based segmentation and matting approaches based on color information only it usually forces a smoothness constraint in the color domain within each segmented region. In the case of color and depth information the same constraint can be adapted to the depth domain, for instance by an exponential function of the color and depth differences such as:

$$V_{smooth}(\alpha(p^i), \alpha(p^n)) = |\alpha(p^i) - \alpha(p^n)| e^{-\frac{\delta(I_C(p^i), I_C(p^n))^2}{2\sigma_c^2}} e^{-\frac{[Z(p^i) - Z(p^n)]^2}{2\sigma_z^2}} \qquad (6.7)$$

where σ_c and σ_z are the standard deviations of the color and depth data noise respectively, $I_C(p^i)$ and $I_C(p^n)$ are the color values of the considered samples and $Z(p^i)$ and $Z(p^n)$ their depth values. The δ function can be any suitable measure of the color difference between the two samples.

The energy functional $U(\alpha)$ of (6.5) can be minimized by efficient graph-cuts optimization algorithms [3] with the methods seen in Section 5.3.

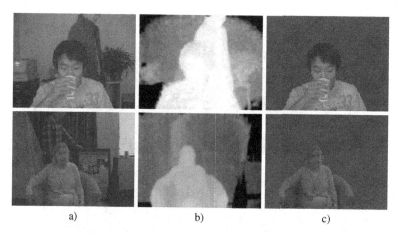

a)   b)   c)

Fig. 6.4 Joint depth and color matting by the graph-cut based method of [26] on two different scenes: a) color image; b) depth map; c) extracted foreground (*courtesy of the authors*).

6.2 Scene segmentation by color and depth data

Scene segmentation is a more general problem than matting, in which the target is the extraction of the different regions of the framed scene. There are a number of approaches to exploit depth data in this task.

The first one is treating the depth map as a grayscale image and applying standard image segmentation techniques to it. This approach gives good results in particular with state of the art segmentation techniques such as [12] and [8].

Depth data are simpler to segment than color images because edges are sharp and well defined and there are no issues due to complex color and texture patterns. However, there are situations that can not be easily solved from depth data alone, such as the case of close or touching objects. Figure 6.5 shows a comparison between standard color segmentation and depth segmentation performed by representing depth maps as images and using standard image segmentation techniques on them (both segmentations types are computed by the state-of-the-art method of [8]). Figure 6.5 confirms that depth segmentation is not affected by complex texture patterns and in general it gives reasonable results, but some critical artifacts remain. For example, note how segmentation by means of depth data alone can not divide the two people that can instead be easily recognized from the color image (compare Figure 6.5b and 6.5d). Moreover, image segmentation techniques do not consider the three dimensional structure behind depth data with various consequent artifacts. For instance, long uniform structures spreading across a wide range of depth values are typically divided in several regions instead of being recognized as a single object (e.g., the slanted wall in Figure 6.5d).

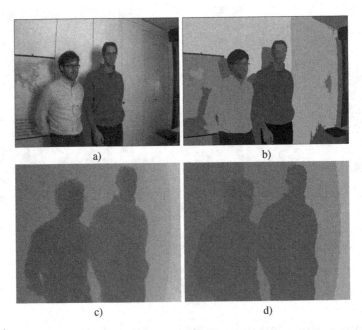

Fig. 6.5 Comparison of color segmentation and depth data segmentation: a) color image acquired by the standard camera; b) segmentation of the color image by [8]; c) depth map acquired by a ToF camera; d) segmentation of the depth map by [8].

Better results can be achieved by taking into account the 3D structure inherent depth data in order to take advantage of 3D clues for segmentation. Furthermore, the color information coming from the ToF amplitude image[1] or from an associated color camera can be used to improve segmentation accuracy. A way of leveraging on the 3D structure of depth data is representing depth data as a 3D point cloud and then applying segmentation techniques developed for this kind of data. Man-made objects usually feature sets of planar surfaces and the recognition of the different planes of the acquired point cloud is a possible depth data segmentation approach [13]. A simple solution to locate the scene planes is to compute the 3D vectors representing the surface normal at each point and then cluster the 3D points on the basis of their normals by a clustering technique, e.g., K-means or mean-shift clustering [5]. The limit of this approach is that planes with the same orientation but different position are assigned to the same cluster. Different planes with the same orientation can be separated in a subsequent processing step from the distance of the planes from the depth camera.

Segmentation based on joint color and depth information usually outperforms segmentations based on color or depth information only in many applications, as shown in the example of Figure 6.6. It can be implemented in two basic ways. The first way is by performing two independent segmentations one based on color only and the other based on depth only, and then combining together the results according to some suitable criterion. For example in [4] the two segmentations are fused together in an iterative cooperative region merging process.

The second way is by exploiting the two clues at the same time, by selecting image segmentation methods and extending them to include depth data, as seen in video matting. An example of this second way of doing is described next. Many image segmentation methods [8] represent each image pixel as a 5D vector (with three components corresponding to color and the remaining two corresponding to the 2D spatial pixel coordinates) and cluster such vectors by state-of-the-art clustering techniques. This approach can be extended in order to handle both depth and color data. The basic idea, used for example in [2] and [10][2], is replacing the 2D coordinates of the image pixels with the corresponding 3D coordinates of the depth data and associating to each pixel a 6D vector instead of a 5D vector as in image segmentation. The three vector components representing geometry need to be expressed in a consistent way and the three dimensional point coordinates $\mathbf{P}^i = [x^i, y^i, z^i]^T$ ensure this since they belong to the same space. It is also possible to use the 2D pixel coordinates $[u^i, v^i]^T$ together with the corresponding depth value z^i, but this requires a further normalization between vector $[u^i, v^i]^T$ and z^i since they lie in different spaces. It is also necessary to represent color in a uniform color space to ensure the consistency of the three color components, and the CIELab or CIELuv color spaces [7] are the most common choices. In any case, geometry and color information lie in completely different spaces, and they must be normalized for a consistent representation. Assuming the usage of the CIELuv color space and the 3D spatial coordinates (x, y, z), each pixel p^i can be associated to a 6D vector

---

[1] In the case of Kinect[TM], the amplitude image is not very informative about the scene color, since it is dominated by the projected pattern.

[2] This work deals with depth data coming from a stereo vision system, but the approach can be easily fitted to ToF and Kinect[TM] data.

    a)   b)   c)   d)   e)

Fig. 6.6 Segmentation based on joint color and depth data information vs. segmentation based on color or depth information only: a) color image acquired by the video camera; b) depth map acquired by the ToF camera; c) segmentation based on color information only; d) segmentation based on depth data only; e) segmentation based on joint color and depth data information.

$$V^i = \left[\frac{L(p^i)}{n_c}, \frac{u(p^i)}{n_c}, \frac{v(p^i)}{n_c}, \frac{x^i}{n_g}, \frac{y^i}{n_g}, \frac{z^i}{n_g}\right]^T \qquad (6.8)$$

with $L(p^i)$, $u(p^i)$ and $v(p^i)$ the L, u and v components of the CIELuv color space at p^i. The normalization factors n_c and n_g are critical for a proper comparison of color and depth data and for balancing their mutual relevance. A possible solution [10] is to normalize both color and depth with respect to their standard deviations σ_c and σ_g. As expected, it has been experimentally found that the segmentation performances strongly depend on the weights between the two types of clues. For this reason, it is practical to deploy a further weighting term λ directly controlling the relevance of depth information, which for segmentation purposes is usually more reliable. The 6D vectors V^i of (6.8) in this case can be rewritten as:

$$V^i = \left[\frac{L(p^i)}{\sigma_c}, \frac{u(p^i)}{\sigma_c}, \frac{v(p^i)}{\sigma_c}, \lambda\frac{x^i}{\sigma_g}, \lambda\frac{y^i}{\sigma_g}, \lambda\frac{z^i}{\sigma_g}\right]^T \qquad (6.9)$$

The proper settings of the weighting factor λ remains critical and its optimal value depends on the scene characteristics. Figure 6.7 shows an example of how the segmentation results depend on the value of λ. Any state-of-the-art clustering technique, such as mean-shift [8], [5] or spectral clustering [21], can be used to cluster the set of vectors (6.9) in order to segment the scene.

A possible variant of this approach is to exploit the derivatives or gradients of depth data [24]. This is similar to using the surface normals associated to depth samples in [5] and it has the advantage that it can discriminate close surfaces with different orientations. Unfortunately the gradients are rather sensible to noise and distant surfaces with the same orientation can not be distinguished in this way.

It is finally worth noting how the just introduced clustering-based segmentation approaches can be applied to any general multi-channel representation, i.e., it is possible not only to consider depth or 3D information together with color, but also any other type of further information provided by Kinect™ or ToF cameras, which simply become further components of the multi-dimensional vectors used in the clustering process.

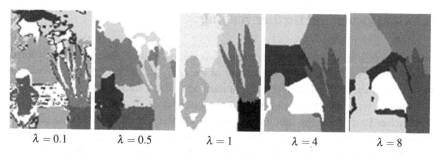

$\lambda = 0.1$ $\qquad$ $\lambda = 0.5$ $\qquad$ $\lambda = 1$ $\qquad$ $\lambda = 4$ $\qquad$ $\lambda = 8$

Fig. 6.7 Segmentation of the scene of Figure 6.6 for different values of the λ parameter in (6.9).

There are two basic approaches for considering multiple clues in segmentation in agreement with the rationale previously followed to implement segmentation based on joint color and depth information. The first approach is to separately consider each clue as proposed in [16], i.e., the scene is first segmented upon each specific clue and the results of all such segmentations are then combined together. The second approach is to consider all the clues at once, for instance in the case of a ToF camera by extending the multi-dimensional representation (6.9) to further components relative to amplitude A_T or intensity B_T and then by clustering the resulting multi-dimensional vectors. Note how the values of the intensity and amplitude A_T and B_T depend on the distance of the acquired points, that is, their values are greater for closer points. This is a rather interesting property for segmentation purposes since it allows to disambiguate objects of the same color at different distances, which is a critical issue for color-based segmentation techniques (e.g., consider the the case of an object in front of a background of similar color).

A critical issue in segmentating ToF data registered with color images is their different resolutions. Kinect^{TM} data have higher resolution than ToF cameras, but still low with respect to that of current color cameras. All the methods presented in this chapter assume the availability of a depth map and a color image of the same resolution. This can be obtained either by the methods described in Chapter 5 in order to have high resolution depth maps or by just subsampling the color image to the (low) resolution of the depth map.

There are also segmentation techniques explicitly handling the different resolution issue. For instance the method of [11] performs first a multi-resolution over-segmentation of the color image and then tries to fit the sparse depth samples inside each segment to a model (e.g., planar or quadratic) of the surface shape . The segments that minimize the fitting error are then selected and used to build an initial segmentation. A region growing and merging strategy also exploiting the sparse depth samples is then used for the final segmentation.

6.3 Conclusion and further reading

Image segmentation is a classical image processing and computer vision problem treated by a vast literature. For a review of the early image segmentation methods see for instance [17], while more recent methods can be found for example in [23]. Among the wide variety of proposed techniques let us recall the clustering based approaches (e.g., [8]), the methods based on graph representations (e.g., [21] and [12]), and the ones based on active contours, watersheds and region splitting and merging. All these methods have been originally developed for color images but can easily be applied to depth maps.

The data acquired by ToF cameras and by Kinect^{TM} can be represented as colored point clouds, therefore current methods developed in the 3D scanning and photogrammetry fields for colored point clouds segmentation, e.g. [19] and [20], can also be adapted to this task.

Finally stereo vision systems are one of the most common sources of depth information and some recent works, e.g. [10], [14] and [15] use depth data from stereo in order to improve segmentation results or try to jointly solve the two problems of stereo depth estimation and segmentation. It is clearly possible to replace the depth data estimated by stereo vision with depth data obtained by ToF cameras or Kinect^{TM} and reuse a large part of the ideas and techniques presented in these works for joint color and depth segmentation.

References

1. O. Arif, W. Daley, P.A. Vela, J. Teizer, and J. Stewart. Visual tracking and segmentation using time-of-flight sensor. In *Image Processing (ICIP), 2010 17th IEEE International Conference on*, pages 2241 –2244, sept. 2010.
2. A. Bleiweiss and M. Werman. Fusing time-of-flight depth and color for real-time segmentation and tracking. In *Proceedings of DAGM 2009 Workshop on Dynamic 3D Imaging*, pages 58–69, 2009.
3. Y. Boykov, O. Veksler, and R. Zabih. Fast approximate energy minimization via graph cuts. *IEEE Transactions on Pattern Analysis and Machine Intelligence*, 23:1222–1239, 2001.
4. F. Calderero and F. Marques. Hierarchical fusion of color and depth information at partition level by cooperative region merging. In *Proc. IEEE Int. Conf. Acoustics, Speech and Signal Processing ICASSP 2009*, pages 973–976, 2009.
5. Y. Cheng. Mean shift, mode seeking, and clustering. *Pattern Analysis and Machine Intelligence, IEEE Transactions on*, 17(8):790 –799, aug 1995.
6. Y. Y. Chuang, B. Curless, D. H. Salesin, and R. Szeliski. A bayesian approach to digital matting. *Computer Vision and Pattern Recognition, IEEE Computer Society Conference on*, 2:264, 2001.
7. Recommendations on uniform color spaces, color difference equations, psychometric color terms. Supplement No.2 to CIE publication No. 15 (E.-1.3.1) 1971/(TC-1.3.), 1978.
8. D. Comaniciu and P. Meer. Mean shift: a robust approach toward feature space analysis. *IEEE Transactions on Pattern Analysis and Machine Intelligence*, 2002.
9. R. Crabb, C. Tracey, A. Puranik, and J. Davis. Real-time foreground segmentation via range and color imaging. In *Computer Vision and Pattern Recognition Workshops, 2008. CVPRW '08. IEEE Computer Society Conference on*, pages 1 –5, june 2008.

10. C. Dal Mutto, P. Zanuttigh, and G.M. Cortelazzo. Scene segmentation assisted by stereo vision. In *Proceedings of 3DIMPVT 2011*, Hangzhou, China, May 2011.

11. B. Dellen, G. Alenyá, S. Foix, and C. Torras. Segmenting color images into surface patches by exploiting sparse depth data. In *Proc. of Winter Vision Meeting: Workshop on Applications of Computer Vision*, pages 591–598, 2011.

12. P.F. Felzenszwalb and D.P. Huttenlocher. Efficient graph-based image segmentation. *International Journal of Computer Vision*, 2004.

13. D. Holz, S. Holzer, R. Bogdan Rusu, and S. Behnke. Real-time plane segmentation using rgb-d cameras. In *Proceedings of the 15th RoboCup International Symposium*, Istanbul, Turkey, July 2011.

14. V. Kolmogorov, A. Criminisi, A. Blake, G. Cross, and C. Rother. Bi-layer segmentation of binocular stereo video. In *Computer Vision and Pattern Recognition, 2005. CVPR 2005. IEEE Computer Society Conference on*, volume 2, page 1186 vol. 2, june 2005.

15. L. Ladicky, P. Sturgess, C. Russell, S. Sengupta, Y. Bastanlar, W. Clocksin, and P. Torr. Joint optimisation for object class segmentation and dense stereo reconstruction. volume BMVC special award issue, pages 1–12, 2011.

16. J. Leens, S. Pirard, O. Barnich, M. Van Droogenbroeck, and J.-M. Wagner. Combining color, depth, and motion for video segmentation. In *ICVS'09*, pages 104–113, 2009.

17. N.R. Pal and S.K. Pal. A review on image segmentation techniques. *Pattern Recognition*, 26(9):1277 – 1294, 1993.

18. T. Porter and T. Duff. Compositing digital images. In *Proceedings of the 11th annual conference on Computer graphics and interactive techniques*, SIGGRAPH '84, pages 253–259, New York, NY, USA, 1984. ACM.

19. T. Rabbani, F. Van Den Heuvel, and G. Vosselmann. Segmentation of point clouds using smoothness constraint. In *International Archives of Photogrammetry, Remote Sensing and Spatial Information Sciences*, volume 36, Dresden, Germany, September 2006.

20. R. Schnabel, R. Wahl, and R. Klein. Efficient ransac for point-cloud shape detection. *Computer Graphics Forum*, 26(2):214–226, 2007.

21. J. Shi and J. Malik. Normalized cuts and image segmentation. *IEEE Transactions on Pattern Analysis and Machine Intelligence*, 2000.

22. J. Sun, J. Jia, C. Tang, and H. Shum. Poisson matting. *ACM Trans. Graph.*, 23:315–321, August 2004.

23. R. Szeliski. *Computer Vision: Algorithms and Applications*. Springer, New York, 2010.

24. M. Wallenberg, M. Felsberg, P. Forssen, and B. Dellen. Channel coding for joint colour and depth segmentation. In *Lecture Notes in Computer Science (Proceedings of the 33rd Annual Symposium of the German Association for Pattern Recognition)*, volume 6835, pages 306–315. Springer, 2011.

25. L. Wang, M. Gong, C. Zhang, R. Yang, C. Zhang, and Y.-H. Yang. Automatic real-time video matting using time-of-flight camera and multichannel poisson equations. *International Journal of Computer Vision*, pages 1–18, 2011.

26. L. Wang, C. Zhang, R. Yang, and C. Zhang. Tofcut: Towards robust real-time foreground extraction using time-of-flight camera. In *Proceedings of 3DPVT 2010*, Paris, France, May 2010.

27. O. Wang, J. Finger, Q. Yang, J. Davis, and R. Yang. Automatic natural video matting with depth. In *Computer Graphics and Applications, 2007. PG '07. 15th Pacific Conference on*, pages 469 –472, nov. 2007.

Chapter 7
Conclusions

This book offers a number of points of interest. The first is a gentle introduction to the operation of CW ToF cameras and of the Kinect^{TM}, the two currently most popular imaging instruments for dynamic scenes capture. The presentation is general enough to apply to the foreseeable evolution of their technologies. This is particularly true for the Kinect^{TM} since Microsoft is very active both in the field of light coding systems, to which the first generation Kinect^{TM} belongs, and in the field of ToF cameras.

The second part of the book focuses on how to get the most from current depth cameras data assisted by one or two color cameras. This requires effective calibration procedures, treated in Chapter 4, and suitable techniques for depth super-resolution and data fusion, treated in Chapter 5. The deterministic super-resolution approaches of Chapter 5 are of particular practical interest, since they can deliver high resolution depth maps with added color information by inexpensive set-up and fast algorithms.

The fusion of depth data obtained by a ToF camera or a Kinect^{TM} with depth data provided by a stereo vision system opens a very wide area (as wide as the literature on stereo algorithms) with intriguing possibilities. Indeed, the synergies between stereo vision systems and ToF cameras or Kinect^{TM} in principle may overcome the various limitations of the single technologies, provided one finds the right recipes for combining stereo data and ToF or Kinect^{TM} data. Chapter 5 offers the instruments for exploring this arena.

This book introduces matting and scene segmentation as application examples of the practical impact of the presented tools and notions, since a proper usage of depth together with color information can bring interesting conceptual contributions to this field together with segmentation tools of significant practical interest. Many other applications could have been considered but were not included for editorial space reasons.

It is worth pointing out a couple of topics which could not be considered in this book, since they are likely to acquire relevance in the near future, namely the study of ToF cameras as MIMO telecommunications systems and the metrological analysis of the ToF cameras and Kinect^{TM} data characteristics and the usage of ToF cameras and of the Kinect^{TM} in body tracking and gesture recognition.